THE ROBOTICS PROGRAM

A How-to-Guide for Physician Leaders on Starting Up a Successful Program

Terrence J. Loftus MD, MBA

The Robotics Program: A How-to-Guide for Physician Leaders on Starting Up a Successful Program

Copyright © 2016

Terrence J. Loftus

ALL RIGHTS RESERVED

No portion of this publication may be reproduced, stored in any electronic system, or transmitted in any form or by any means, electronic, mechanical, photocopy, recording, or otherwise, without written permission from the author. Brief quotations may be used in literary reviews.

ISBN: 978-1-365-02193-0

FOR INFORMATION CONTACT:

Terrence J. Loftus

Loftus Health

60 East Rio Salado Pkwy, Suite 9033

Tempe, AZ. 85181

www.LoftusHealth.com

Dedication

This book is dedicated to my wife, Jennifer Loftus, for her continued faith in me, support throughout our marriage and especially during the writing of this book and for encouraging me to scratch that entrepreneurial itch.

ACKNOWLEDGEMENTS

I want to personally thank the following individuals for reviewing this manuscript and providing critical feedback and much appreciated encouragement and support along the way.

Mark Hausmann, MD

William Hope, MD

Sachin Kukreja, MD

DISCLOSURES

Dr. Loftus has been a speaker for Intuitive Surgical, Inc.

TABLE OF **CONTENTS**

CHAPTER 1: SEVEN PILLARS	2
CHAPTER 2: PURPOSE	10
CHAPTER 3: ENGAGEMENT	18
CHAPTER 4: COMMUNICATION	34
CHAPTER 5: INFRASTRUCTURE	42
CHAPTER 6: ACCOUNTABILITY	54
CHAPTER 7: LEADERSHIP	62
CHAPTER 8: PERFORMANCE IMPROVEMENT	72
APPENDIX: COMMITTEE CHECKLIST	88

CHAPTER 1
SEVEN PILLARS

Several years ago I began a new role as a physician executive for a large integrated delivery network. My previous role was as a Chief Medical Officer of a large community hospital and before that as an Acute Care Surgeon at a Level 1 Trauma Center. The new role was a corporate position, and my title was Medical Director of Surgical Services and Clinical Resources. One of the primary functions for the system's surgical services was to oversee clinical quality and patient safety. This involved guiding the development and implementation of clinical practices across the system. The other primary function was to work along side, and advise supply chain services, especially with regards to surgical services.

On my first day of work at the corporate office, one of the members of our senior management group stopped me in the hall and asked, "Terry, what do you know about robots?"

"Not much", was my reply.

"Do you have a strong belief, either way, regarding robotics? He pressed.

"Not really", I said. "I'm pretty much an agnostic."

"Perfect", he said, "we would like you to learn as much as you can".

And so began my journey. His real concern was the system invested millions of dollars in robotic technology, and was still not sure if it was a good investment. They wanted someone who was a surgeon who could provide insight into this investment, and guidance for the system. They received plenty of opinions from surgeons who either supported or didn't support the use of robotics. What they needed was someone who was relatively neutral to the technology, and was willing to do a deep dive into the vast amounts of data the system collected on robotics. The hope was that mining the data of twenty robotic systems in thirteen facilities would reveal new insights regarding the technology. I'm a bit of a data geek and I love a great mystery, so I was perfect for this adventure. What I hope to convey to you in this book is what I learned along the way.

Before we delve any further, I want to say a word about who the audience for this book is. It is primarily for people who want to develop a Robotics Program for a hospital or system. It is also for those who already have a program in place and are looking for ways to improve it. While it doesn't matter which type of robotics systems you are using, this program was primarily developed around the Intuitive Surgical system. By the end of 2015 there were almost 3,600 Intuitive Surgical da Vinci systems installed worldwide[1]. Two-thirds were in the United States and one out of every six systems was in Europe. There is a good chance you are working in one of these facilities. Regarding the evidence, this book will not go into an exhaustive review of the clinical literature regarding robotic assisted surgery. In addition, this book is not a replacement for other programs such as: Intuitive Surgical's Genesis Training Program, the ECRI Institute's Robotic Surgery Planning Program, The Center for

---

[1] Intuitive Surgical FAQs, http://phx.corporate-ir.net/phoenix.zhtml?c=122359&p=irol-faq. Accessed March 9, 2016.

Applied Value Analysis (CAVU) program for robotics or the program from the Texas Institute for Robotic Surgery. This book is meant to supplement these excellent programs. You are reading this book because you wish to expand your knowledge of how to build a successful Robotics Program. You are also reading this book because you may still have some doubt, and want to know what would transform someone from a robotics agnostic to a robotics believer.

As you may have already determined, I am no longer agnostic to the technology. I believe it has a role in the discipline of surgery. I also believe its role is slowly evolving as the technology improves, and the clinical evidence for its best use is slowly being revealed in the literature and in the proving grounds of the operating room. As it turns out, it is not as simple as buying a robot and setting it up in an operating room. Yes, it is a tool, and like all tools it works best under specific conditions. This book is about what programmatic elements need to be in place to get the most out of this tool for optimal utilization, cost and clinical outcomes. This book is presenting a hypothesis. The hypothesis is: A structured Robotics Program with specific, defined elements produces better quality at a lower cost than a relatively unstructured program. The good news is there is some support for this hypothesis, although much of it is built on limited data and anecdote. We must start somewhere and this is as good a place to start as any. In fact, much of what we practice in medicine has a similar beginning. So back to the start of the story, and how I went from agnostic to believer.

I love data. In my role I had access to an unprecedented amount of data about robotics. It was everything the system collected. It included financial, utilization, supply cost, equipment cost, and clinical outcome data. We even had data provided by our vendors. I could slice and dice it any way I needed. I reviewed data for the system, region, facility, specialty, service line, case type and down to the individual surgeon. The data was from multiple years (2010 – 2015), so I could see trends in all of these various categories. The data could be risk-adjusted and bench-marked to peer facilities across the United States. From all of this data, and my experience developing Robotics Programs, I started to see certain patterns emerge. When I began describing these patterns in lectures across the United States, I encountered other people, in other healthcare systems, who were seeing similar patterns. This book is a compendium of what I learned in the process. We are going to begin at the sixty-thousand-foot "system level" and bring it down to the dry, sometimes very mundane, "operational level". So to begin, what is a Robotics Program?

A Robotics Program is a facility or system based program, which is a comprehensive, multidisciplinary approach to implementing, supporting and utilizing robotically assisted technology for surgical procedures. The purpose of the Robotics Program is to improve the quality outcomes for patients, optimize the utilization of the robots, work to decrease the total cost of care of surgery and enhance the work place experience of the staff involved in providing care to patients undergoing robotic assisted surgery. A Robotics Program is the overarching structure, where as, a Robotics Committee is the formal Executive arm which provides oversight in determining how that structure is integrated into the system and how it performs. In the remainder of this book we will discuss the specific features of a successful Robotics Program and especially successful Robotics Committees. In the remainder of this chapter we will focus on what I refer to as the Seven Pillars of successful programs. These are higher level elements that all successful programs build into their structure. To remember the Seven Pillars, think of the mnemonic "**SPECIAL PI**". This stands for, **S**even pillars, **P**urpose, **E**ngagement, **C**ommunication, **I**nfrastructure, **A**ccountability, **L**eadership and **P**erformance **I**mprovement. Chapters two through eight will provide a description of each element in detail. The appendix is a checklist for how to start-up a successful program and specifically a successful Robotics Committee. The following is a summary of the seven pillars.

THE SEVEN PILLARS

1) **PURPOSE:** There are key elements to any successful Robotics Program. It begins with having a purpose. Everyone on your team must understand the "why". In chapter 2, we discuss the why in detail. In summary, it is because a formal organized approach to delivering robotic assisted surgery produces better outcomes at a lower cost compared to not having a program.

2) **ENGAGEMENT:** Engagement has become a buzz word in the healthcare industry. More specifically hospital administrators and physician leaders want "physician engagement". For many it has become a quest for the Holy Grail. The unstated belief is, "If we only had physician engagement, then we could solve all of our problems in healthcare." There is no doubt that physician engagement is important for change management in healthcare. Before we get to this place though, we need to set up our program to become engaging. Chapter 3 will discuss this in greater detail. In summary, the first step in this process is to stop doing those behaviors that disengage people and move on to those behaviors that create an engaging program.

3) **COMMUNICATION:** Programs must communicate to their members, leadership, patients and other stakeholders. It is impossible to educate and inform people without a consistent process for communicating to them. Communication must be a two-way process. A program must not only communicate to its stakeholders but its stakeholders must have a mechanism to communicate to the program's leadership. Chapter 4 covers more specific information on how to communicate effectively. In summary, effective communication increases stakeholder support. If you want to be successful, then you must communicate effectively.

4) **INFRASTRUCTURE:** All programs must have a well developed infrastructure. A program's infrastructure is the foundation on which the program will develop and sustain itself. This is often

considered the boring work of building a program and therefore nonessential. Unfortunately, many organizations overlook this important element. In chapter 5 we will discuss what exactly infrastructure is. In summary, it is all the detailed, value-added process steps that increase the likelihood of a highly functioning program.

5) **ACCOUNTABILITY:** Without accountability all programs eventually fail. Robotics Programs perform best when the accountability is integrated into the structure and governance of the program. We will talk about the pros and cons of different models in Chapter 6. In summary, someone or some group must be accountable for the following.
   a. Implementing the program
   b. Managing the program
   c. Establishing goals for the program
   d. Creating action plans to achieve goals
   e. Reporting results to Medical Staff and Administrative leadership

6) **LEADERSHIP:** In order to successfully navigate the complexity of our modern healthcare system, there must be an individual or dyad at the helm of the program. When we think of robotics we tend to focus on the technology. While the technology is an important part of any program, programs are implemented, comprised of, managed and overseen by people. People want and expect good leadership as a part of any program. We've found that a dyad leadership structure is particularly effective. This is typically made up of a physician leader and a nursing leader in the hospital setting. Chapter 7 will discuss how leadership will impact your program. In summary, all stakeholders will have varying expectations of you as a leader. Understanding what each stakeholder will expect from you will improve your ability to lead.

7) **PERFORMANCE IMPROVEMENT:** Performance improvement answers the question of how we are going to improve value for

patients and the people who provide their care. In addition, it is very difficult, if not impossible, for a program to improve without data. Data is the backbone of performance improvement. Whenever possible, the data should focus on quality, utilization, cost and patient/provider experience. While some data may not be accessible to a program, this should not prevent any program from developing a data source. Performance improvement has been done with pencil and paper long before computers, databases and statistical software. While this may seem primitive, based on modern capabilities within most healthcare systems, it is important to note that data comes in many forms, and data analysis is only limited by our imagination and willingness to be creative about how we collect and interpret data. Chapter 8 will provide a more detailed description of this. In summary, you manage what you measure. If you want to improve, then you will always want to know, compared to what?

CHAPTER 2
PURPOSE

Since 2000 the annual growth in robotic assisted surgery has increased tremendously. Between 2009 – 2014 the number of procedures performed on the da Vinci platform from Intuitive Surgical has tripled from approximately 40,000 to 120,000[2]. It is not just the number of procedures that is growing, but also the volume of literature. In 2014 there were approximately 1400 articles published in peer reviewed journals on this topic. This growth in procedures along with the advancements in the robotic technology and knowledge has created an interesting challenge for healthcare systems. How do we manage this?

In the past when a hospital purchased a piece of equipment for the operating room, there was an understanding that once the surgeons and nursing staff were trained on the equipment, that was all that was needed. Robotics changed that. In addition to the procedural growth and knowledge came an explosion in data. We began to learn not just about the strengths and weaknesses of the technology but also, more

---

[2] Intuitive Surgical, Inc. 2014 Annual Report

importantly, about the context in which the technology was being deployed. As it turns out, people, process and culture are just as important as the technology for determining outcomes and therefore the success of Robotics Programs. There is a tendency to blame the technology, when more often than not, the issue is one of people, process or culture. More specifically, the issue of how the people, process and technology are organized and work together as a program is what defines the culture. How you describe your Robotics Program is really how you describe your robotics culture. Peter Drucker, the founder of modern management[3], is reported to once say, "culture eats strategy for breakfast." Strategize all you want. Keep in mind though, building an effective culture begins with having a purpose.

PROGRAMS MATTER

During the process of organizing our system Robotics Committee, we collected outcomes data for each of the facilities. There were obvious differences between facilities, and the first hypothesis was that these were facilities with greater experience and volume. This wasn't always the case. High-volume surgeons were distributed across all the facilities. Something else was driving outcomes at each facility. About this time, we were collecting data on the organizational structure of Robotics Committees based at each hospital. (The chapter on Infrastructure will describe these data elements in greater detail.) Our goal was for the System Oversight Committee to get a better understanding of how each facility approached robotics. What we found was that some facilities were very organized and highly functional. Other facilities were either poorly organized or did not have a Robotics Committee. We then documented all of the organizational elements these committees had. We were then able to grade each of the facilities based on the number of organizational elements each had. We defined highly structured programs as facilities that consistently used more than 75% of these elements.

---

[3] https://en.wikipedia.org/wiki/Peter_Drucker

Less structured facilities used less than 75%. We then compared the outcomes (complications and cost) of highly structured to less structured programs. The results were very interesting.

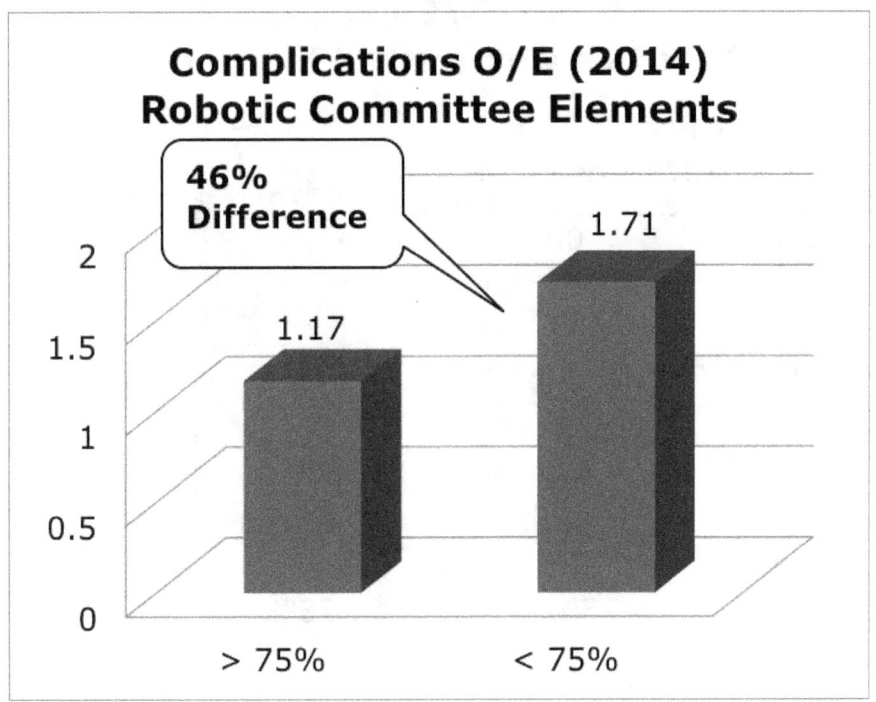

Chart 1: Comparison of the observed to expected complications between highly structured Robotics Programs (>75%) and less structured Robotics Programs (<75%).

Chart 1 shows the comparison of complications between the highly structured programs (>75%) to the less structured programs (<75%). On the vertical access is the observed to expected ratio for complications at each facility using the select data from Premier[4]. These are risk-adjusted and bench-marked data. Less structured Robotics Programs had a complication rate that was 46% greater than Robotics Programs that were highly structured. When the same analysis was performed for cost there was also a difference. (See Chart 2) Less structured Robotics Programs were found to have costs that were 37% higher compared to highly structured programs.

---

[4] Premier Quality Advisor™ 2014. (Unpublished data)

Needless to say, our organization decided to support the development of highly structured Robotics Programs across the system.

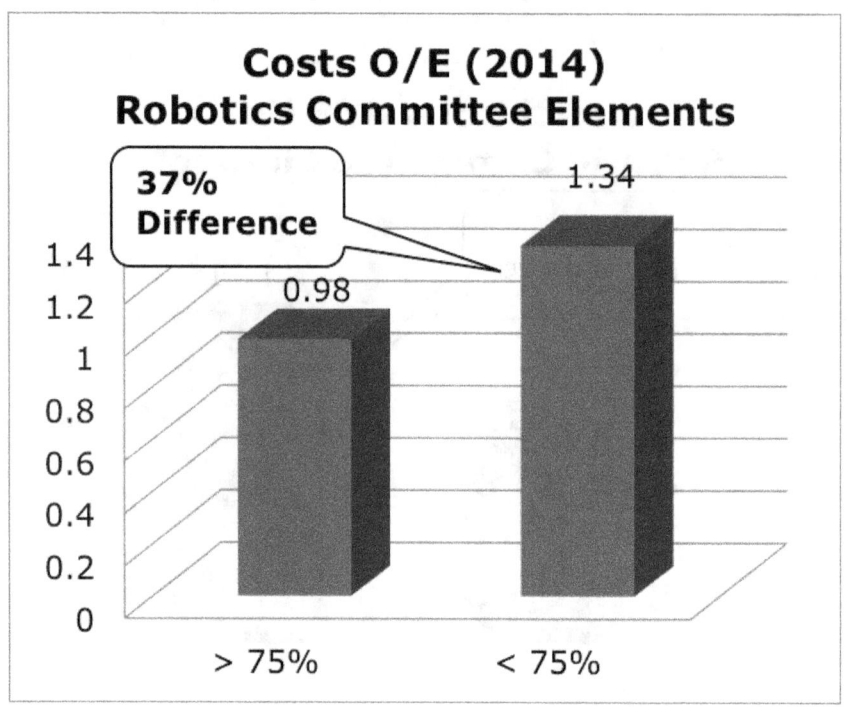

Chart 2: Comparison of the observed to expected cost between highly structured Robotics Programs (>75%) and less structured Robotics Programs (<75%).

VOLUME MATTERS

Sitting in our System's Robotics Oversight Committee meeting one evening, the question of volume was discussed. The question was, "Do high-volume surgeons have better outcomes than low-volume surgeons"? We had access to risk-adjusted and bench-marked data from the Premier database and reviewed the analysis.[5] High-volume surgeons were defined as those who were in the top 20th percentile in terms of annual volume. Low-volume surgeons were in the remaining 80th percentile. The risk of complications was 36-38% greater for low-volume surgeons compared to high-volume surgeons in 2013 and 2014. (See Chart 3) Costs were 10- 20% higher for low-volume

---

[5] Premier Quality Advisor™, 2014. (Unpublished data)

surgeons compared to high-volume surgeons. (See Chart 4) Similar observations were made for length of stay (11% higher) and readmissions (66% higher).

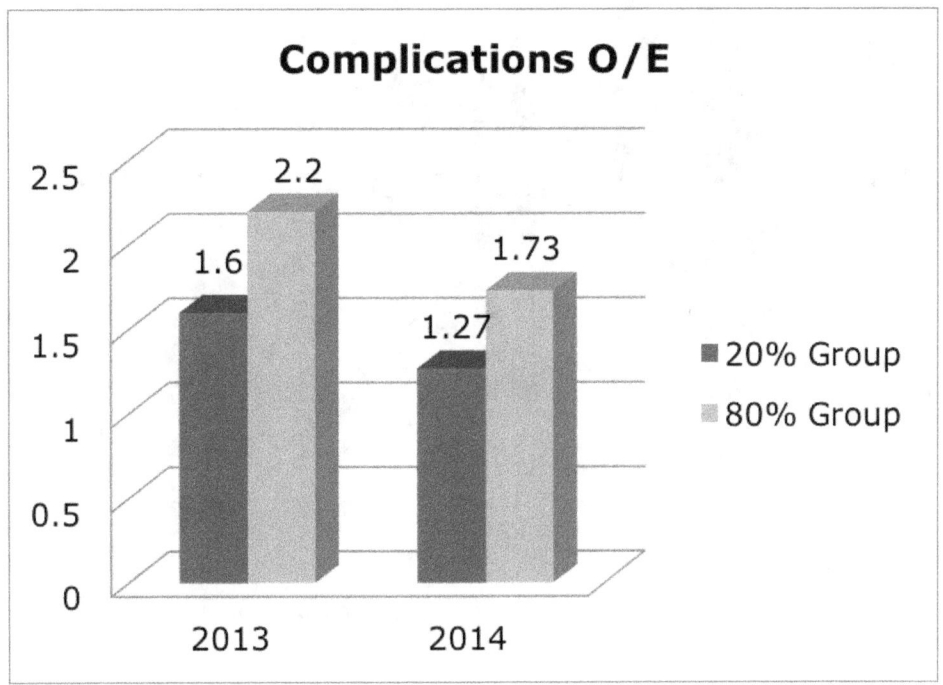

Chart 3: Comparison of the observed to expected complications between high volume surgeons (top 20%ile) and low volume surgeons (bottom 80%ile).

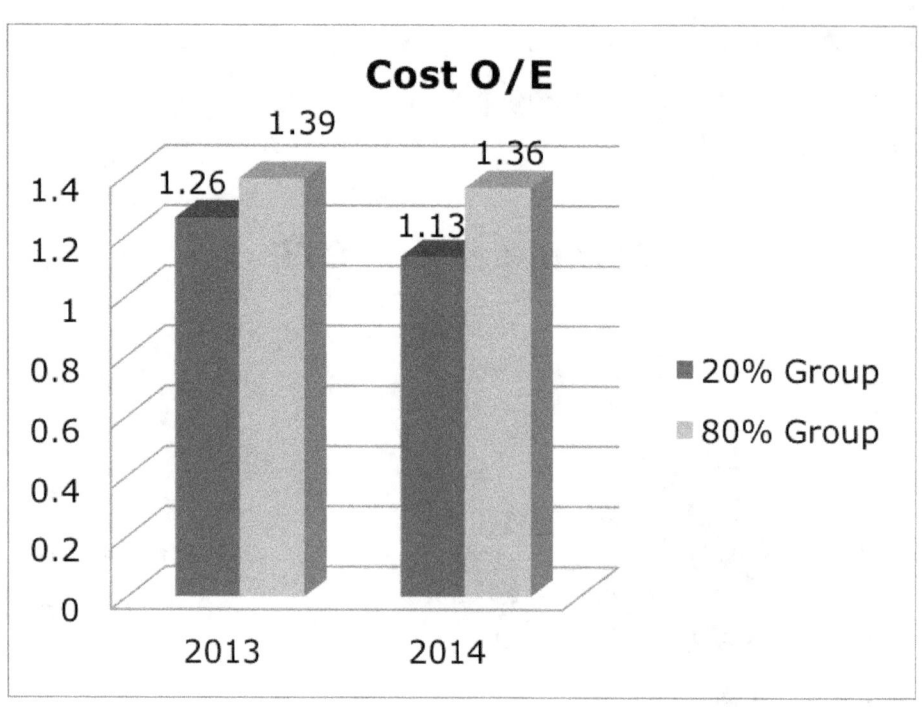

Chart 4: Comparison of the observed to expected cost between high volume surgeons (top 20%ile) and low volume surgeons (bottom 80%ile).

Of note, the break point for number of annual cases between the two groups was 24 cases/year, or two cases/month on average. As you can imagine, the averages do not tell the whole story.

There was also another interesting observation best illustrated by a Pareto chart. (See Chart 5) On the horizontal axis are individual surgeons ranked in order (left to right) from lowest volume to highest volume. On the vertical axis is the total number of cases performed from 2012 – 2014. The chart shows a small number of surgeons perform a large number of procedures, and a larger number of surgeons perform a smaller number of procedures. Average for 3 years is 72 cases, however the chart shows the break point for the group is closer to 150 cases or about 50 cases per year. In 2013 the busiest surgeons were performing about 60 cases per year and the least busy surgeons were performing 10 cases per year. By 2014 the gap increased. The busiest surgeons were now performing 80 cases/year. The least busy were only performing 5 cases/year. What we observed is that the cases are migrating to the busier surgeons.

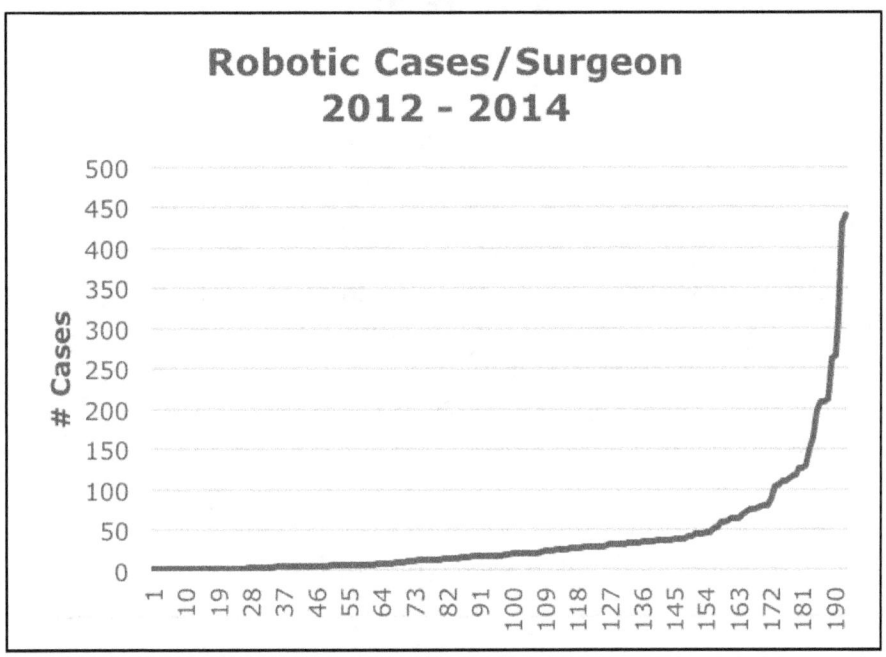

Chart 5: Pareto chart of surgical volume for individual surgeons between 2012 – 2014.

What is less clear from the data is whether highly structured programs support and attract busy surgeons or vice versa. Our data was insufficient to declare one way or another. Just based on observation, these two variables appear to be mutually reinforcing. One last observation to mention before we begin the next chapter.

Productivity Matters

It may seem obvious by now, but highly structured programs with busy surgeons also tend to be highly productive facilities. Chart 6 shows the relationship and trend between productivity and cost.

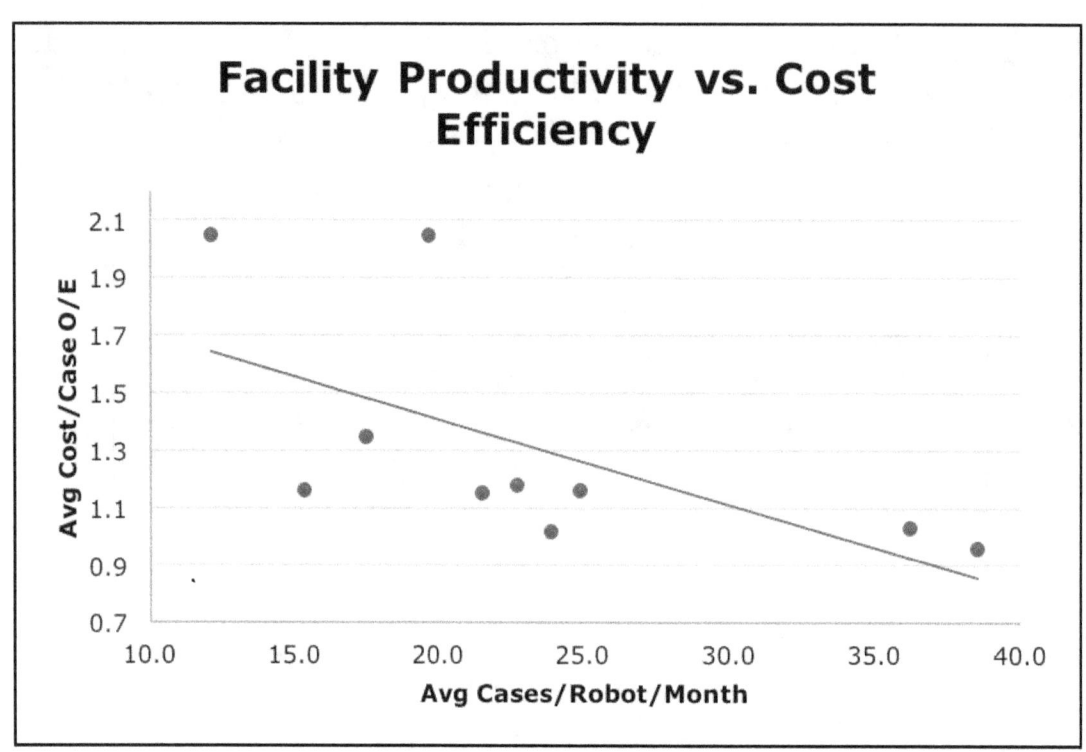

Chart 6: Relationship between the observed to expected cost for facilities and each facilities productivity

The vertical axis shows the observed to expected ratio for cost at each facility. The horizontal axis shows productivity as measured by average number of robotically assisted cases, per robot, per month. While it is difficult to make any firm conclusions on this data, the overall trend line indicates a cost advantage for facilities that are busier.

CHAPTER 3
ENGAGEMENT

Engagement is the buzz word for the 21st century in healthcare. No where is it more commonly applied than in the expression "physician engagement". As discussed previously it is the Holy Grail in healthcare. In this chapter we will discuss an alternative approach to engagement starting with a few introductory recommendations.

The first recommendation is to quit focusing on just physicians. Too often many of us in leadership point to a lack of physician engagement as the sole source of the problems in healthcare. The thought being, if only we had more physician engagement, then we could solve all of these problems. It is a false hope. We do need physicians to help with much of what ails our healthcare systems, but we need to keep in mind that they are not the cause and the sole source of the problems. Healthcare is complicated and has a lot of moving parts, as they say. Physicians alone are not going to solve all of them. Instead of "physician engagement" what we need is engagement. The principals of engagement work for all people, including physicians.

The second recommendation is to recognize that physicians are already engaged. In the past when I was a busy Trauma Surgeon, there was not a day when I did not feel engaged. When I wasn't at work in the operating room, clinic, ward, trauma room or ICU, I was probably still thinking about my practice in some form. I could be talking to a colleague, reading a journal article or just dreaming of what I would do in a particularly challenging case. I felt like I was always engaged, much as I do about my current work. This was true of just about every physician I worked along side. When someone complains and says physicians are not engaged, what they really mean is physicians are not engaged in the same things that person would like them to be engaging. So how do we get people to become more engaged in things that they are not currently engaging? That is the topic of the remainder of this chapter.

DISENGAGEMENT

A colleague mentioned something to me that seemed trivial at the time. He became one of the newest members of a system group that was looking at alternative supplies in the operating room. He seemed to have a particular passion for this topic, and I was looking forward to his participation. He lived quite a distance from where the group was meeting, so we set up a teleconference to allow him and other team members to join in the discussion. He called in once and then we never heard from him again. I ran into him sometime later and asked him why he was no longer participating. He said, "every time I call into that number, I'm put on hold for what seems like forever, so I just stopped calling." That particular group had a lot of members who were chronically late so we tended to start the meeting late. It was not uncommon for the meeting to begin as much as 10 minutes late. When this occurred, we would put the phone on hold. One big mistake leading to another big mistake.

Before we can engage people, we need to stop doing those behaviors that disengage them from the very things we want them to be engaged. In this case, as Chair of the committee, I allowed the meeting to routinely begin later than scheduled. Since we ended up starting later each time it encouraged members who showed up in person to arrive later each time. For our call in line, we would announce very early in the call that we were waiting on some more people to show up for the meeting. Following this we put the phone on hold. If anyone called in after this, they were not always aware of what was happening. There are two big problems with this. The first is by starting the meeting later, I was sending a message to the people who arrived on time. It goes something like this. You (who showed up on time) are not as important as the people who are late. It is never a good thing to insult the people who are prompt and ready to engage. This is an example of behavior that is disengaging. Little did I know, it was also leading to another behavior that was disengaging the people who were calling in. Over time, the membership of that group dwindled. At the time we concluded that the cause of this was a lack of "physician engagement". As you can see, this conclusion was incorrect. We started with an engaged group, and slowly and very unconsciously disengaged them over time. The problem was disengagement or better described as behavior that is disengaging.

There are common attributes to behavior that is disengaging. The first is that is inconsistent with people's expectations, may seem trivial at the time it occurs and appears to the receiving party as inconsiderate. If I tell people the meeting will start promptly at 5:00 PM and it doesn't, then I failed to meet their expectations for being prompt. If I reward others who are late by delaying the start of the meeting, then I am inconsiderate to those who arrived promptly. Many of us run into this scenario on occasion, so most of the time we are willing to excuse this behavior. If this happens repeatedly, then we become intolerant to this behavior. We feel the need to punish the person who is doing this to us. It is also interesting to note that it wasn't just physicians who quit showing up to the meetings. It was most of the members, physicians and non-physicians. This was equal opportunity disengagement at its worse. This is what we will refer to as Category I DBs (disengagement behaviors). If performed repeatedly, then they result in disengagement of just about everyone. Category II DBs are directed at a specific target audience.

We were at a Department of Surgery meeting and the next agenda item called for a report from one of our administrative managers. Sixty seconds into the report I could tell this was not going well. The surgeons were getting restless and the Chair was motioning to everyone to settle down. What could possible be said in under a minute that could so enrage this group. In that short time span, the words or phrases "compliance", "mandatory", "Joint Commission" and "audits" were somehow use in the very first sentence the manager uttered. While these words may seem routine to someone from the hospital administration, they tend to derail many physicians and especially surgeons. This is not the best way to start a conversation. Whatever message the manager meant to communicate was completely lost on this group. The group immediately piled on the manager before she could finish her opening remarks. This led to a thirty-minute rambling, tangential heated discussion on why we even needed the Joint Commission. We never discussed what the manager was attempting to tell us. What was that you ask? Our hand washing rates were improving! The manager came to give them good news and ends up getting verbally abused for a poor choice of words. When she gave the same presentation a week later in our C-Suite meeting she received applause.

Category II DBs are group and sometimes individual specific. We've all encountered a situation when we are completely turned off by another person's behavior, and the people we are with are not impacted by this at all. It is impossible to know exactly what behaviors will disengage specific groups and especially individuals. As we get to know people some things become predictable. For physicians, there are predictable Category II DBs that should be avoided whenever possible. If you can state a message without mentioning or doing a Category II DB, then that is what you must do. In the previous paragraph, the story demonstrates one of the biggest verbal blunders you can do with a physician audience. Physicians value autonomy and independence. They do not like to be reminded that others can and do control their behavior to some degree. There are buzz words that trigger that sense of loss of autonomy and independence. Words like compliance and mandatory will quickly disengage this audience, especially if the discussion begins with these words.

Another way to disengage a physician audience is to begin a discussion with just about anything related to financial concerns. When I was in my surgical training, we were told that we were not to think about the cost of what we were doing and to focus on doing what is right for the patient. It seemed like great advice at the time. Fast forward 25 years and now it is impossible to not consider cost in just about everything we do. Physicians tend not to value the contributions finance has made to healthcare. Like it or not, we are now being exposed to the world of finance on a regular basis. Despite this, we still become disengaged whenever someone leads off a conversation or presentation on cost, and especially cost reduction. It's easy to drift back to a time when we only had to think about what is right for the patient, regardless of cost. We were trained to think this way, and we will always value patient care over finance any day of the week. In business school, the rules are different. One of the first things I learned was that all roads pass through finance. (The second thing was, there is no free lunch.) If you want to get just about anything done in a healthcare system, then you will need to learn how to navigate this road through finance. So how do you have discussions with physicians about financial issues?

Lead conversations with physicians with what they love, and that is patient care and quality. Demonstrating the relationship between improving quality and some topic that I am not interested in hearing, such as cost improvement, is a way to improve the likelihood that I will listen to you. Focusing only on a topic a physician finds disengaging is going to lose them very quickly. So what about those times when you must do something that will inevitably trigger disengagement?

When we know, or even suspect, something we are going to do will lead to disengagement, then we need to consider mitigating behaviors. These are behaviors that mitigate, or reduce the negative effect of the disengaging behaviors. In our previous example, where I put people on hold waiting for other members to arrive, I could have gotten back on the line and reminded everyone that the meeting would be delayed for a few minutes more. In addition, I could have also take the time to remind everyone that in the future, we will not wait for late arriving members, and will begin the meeting promptly. I began this practice later in my career and also included ending meetings on time. This is a reactive type of mitigating behavior. I realized I was behaving in a manner that was disengaging the people our team was dependent on for our success. I learned from it and changed my behavior in future meetings. Getting back to those situations in which we know our behavior is going to be disengaging to our audience. This is a time for a pro-active mitigating behavior.

Supply Chain Services encountered a problem while negotiating for a particular supply that was a known Physician Preference Item (PPI). In order to negotiate a lower cost, we were either going to need to pressure the suppliers to lower their cost or reduce the number of suppliers. By reducing the number of suppliers, we would open market share and the remaining vendors would be more likely to reduce their prices. We knew that eliminating vendors would likely cause some bad feelings among some of the surgeons. Instead of making a decision solely on price, we decided to bring the surgeons in on the discussion through a value analysis team approach. We ran a usage list for the PPI in question and brought together a team of surgeons who used this product category on a regular basis. This team of subject matter experts were able to help us on a contracting strategy which led to over half a million dollars in supply savings per year. This was a win-win for the surgeons and healthcare system. Based on the success of this team, we developed a Value Analysis Program that went on to become highly successful. Not all proactive mitigating behaviors enjoy this kind of success.

Once again Supply Chain found itself in a high tension negotiation. This time the majority of suppliers for a particular type of PPI were willing to meet our price point for the product. One supplier, which happened to have a small group of fiercely loyal supporters, was the lone hold-out. Supply Chain took a proactive approach and met with each of the surgeons who supported the product. While they could understand the system's dilemma, many were unwilling to support changing vendors. We knew we were going to need to walk away from the negotiating table with this particular vendor, as the price they were demanding was off the charts compared to the fair market price range. We developed a time table for the surgeons as to when we would no longer provide that product, and a plan to transition to different products. A couple of surgeons moved some of their business to the competition, however the majority transitioned to an alternative product. Eighteen months later the vendor, after losing all of their market share in a dominant healthcare system in the the market, came back with a revised offer that was consistent with the revised price point for the product, which was now lower. Had they met the original price point, they would not have lost all of that market share, and would currently be priced slightly higher than where they ended up when they returned. While a few surgeons were not happy with the original deal, most were accepting of the process. Having a defined process for this situation is what made the difference.

The single most important thing you can do to engage people is to start by not disengaging them. There are behaviors we do that can disengage just about anyone. Not only do we need to learn how to remember to not do them, we also need to take the extra step and learn how to hardwire this into our infrastructure. So even when we do forget, we can be reminded by one of those hardwired triggers. The second most important thing we can do is change our behavior when we realize it is creating disengagement. It may not always be obvious to us, but when people stop showing up to your meeting, then there is usually something that is either driving them away or is more engaging to them. The third most important thing we can do is to proactively mitigate disengaging behaviors that must be done. When we know something is going to disappoint a stakeholder group in our organization, then it is time to be proactive. If you were them, then how would you like to be treated? Usually this is by being forewarned of any changes, and being provided time and resources to adapt to the change.

ENGAGEMENT

One of the biggest falsehoods I hear repeated is that physicians and surgeons in particular are resistant to change. Nothing could be further from the truth. Yes, they will be assertive. Yes, they will question the need for change. So are physicians' resistant to change? The short answer is no. There has been more change in the last 100 years of medicine than in its entire history. In just the last 25 years, we have experienced an unprecedented amount of change in surgery. Previously most of the abdominal cases were performed in an open fashion. Today, most are performed with a minimally invasive approach. This has led to changes in instruments, equipment, medication use, hospital utilization, etc. The list goes on. During this time, physicians were always changing and always demanding to understand why change was necessary. There is nothing wrong with that. Physicians are really much more change hardy than we like to believe. They are also more engaged than we like to think.

When it comes to change management in a hospital or healthcare system there is an enormous amount of time, energy and resources devoted to physician engagement. The underlying theme of many of these efforts is, how do we get physicians to change their behavior and get them to do what we want them to do? Physicians, as we previously discussed, value autonomy and independence. They are not going to change just because you want them to change. For this sin they are branded as "change resistant", "laggards" and "Luddites". Yet, this is how some of the most change hardy people on the planet are treated. Physicians are engaged, they are just not engaged in the things a typical hospital administrator finds engaging. So how do you achieve physician engagement? You have to know what they already find engaging. How do you learn what a physician finds engaging? Well, for starters you can ask them. When you can't ask them or they are not responding to your question, there is an alternative method. It's not perfect, but it can provide some insight into what they are thinking, and this gets you closer to what they find engaging.

We were in the early phase of developing a Robotics Program for our system and wanted to pull together a group of surgeons who did robotic assisted surgery. We called around the system and received a few names. They were all gynecologists based out of two facilities. It was obvious this was not representative. We wanted broader representation and people who were truly passionate about the technology. There is a fairly simple way to identify these people. We ran a usage list. We identified all the surgeons who performed robotic assisted surgery along with their specialty, number of cases, type of cases, years with robotics privileges and outcomes. The basic list of usage was easy to pull. The outcomes took a little longer, but we were able to now get a list of our busiest surgeons with the best outcomes. We told them we were interested in developing a system robotics team looking at improving outcomes, and asked them if they had any interest in this. Guess what? They were very interested in joining this group. Why? Because they were already engaged in the topics of improving outcomes and robotics and wanted to learn more. Imagine that, instant physician engagement.

Rule number 1: *If you want physician engagement, then start with what they are already engaged in doing.*

We repeated this process multiple times when we developed a Value Analysis Program. This is a program, usually managed through Supply Chain Services, that brings together subject matter experts to advise the healthcare system on equipment, supplies and servicing contracts. Physicians are typically recruited for these value analysis teams. When we first started the program we took advantage of this usage list approach. We could easily see who used certain products and vendors the most. We could also see what types of cases the surgeons were performing most frequently. Surgeons love the tools they use for operating, and if there is going to be any discussion regarding them, they want to be at the table. (Many have learned the hard way, that if you are not at the table, then you are on the table.) Usage lists tell you not only what types of cases interest them, but also what types of tools interest them. When you want to include a discussion on outcomes, then find those physicians who are already getting great outcomes. They are almost always interested in improving. Why? Because that is what they are already engaged in doing.

We were implementing an enhanced recovery after surgery program for bowel surgery across our system, and trying to determine if there was any support for the program. Once again, we turned to the data. We planned on monitoring two key steps in the program. They were early ambulation and early alimentation. We ran a list of surgeons who performed these types of cases and a list of how often these two key steps were being performed by their patients. Overall these steps were being performed only 30-40% of the time on average. What was really interesting was that a majority of the surgeons were ordering early ambulation and early alimentation much more frequently than this. Evidence based medicine supported this practice. The surgeons were aware of the evidence, and thought they were doing it as best they could. Unfortunately, the actual performance and outcomes were not matching this expectation. When we implemented the program we built decision support into the order sets and created a tracking board of these patients to make the individual patients' performance more visible. Not only did the performance of these key steps improve significantly, but is was also associated with a significant improvement in complications and readmission rates[6].

Rule #2: *If you want physician engagement, then help them solve a problem they are already attempting to solve.*

For ten years in a row we endured continuous price increases in a particular product category. So we ran a usage list for this product, and pulled together a group of surgeons with the greatest use. What was particularly troubling about this product was, the hospitals lost money every time this product was used. The amount was substantial, and was creating a risk for the service line. We were at the breaking point, and were not going to be able to support this product and the surgeons anymore to the degree we did in the past. There were a lot of alternative plans discussed, but we wanted to know what the surgeons thought. They were appalled. They were also unaware this whole time how much the product cost, and that the hospitals were losing money every time it was used. With the support of the surgeons, we approached the vendors regarding a significant price decrease. Initially they were not interested and wanted another price increase. They appealed to the surgeons for support. Not one surgeon would support the vendors, and told them so in no uncertain terms. We ended up with the price decrease we needed to sustain the program.

---

[6] Loftus T, Stelton S, Efaw BW, Bloomstone J. A system wide care pathway for enhanced recovery after bowel surgery focusing on alimentation and ambulation reduces complications and readmissions. J. Healthcare Quality. Published online Feb 20 2014. doi: 10.1111/jhq.12068

Rule #3: *If you want physician engagement, then appeal to their sense of fairness and what is right for the sustainability of their service line or even the hospital in which they practice.*

There are going to be times when you must address issues that many physicians will simply not be interested. We ran into this with supply cost in the operating room. Just about every surgeon whose supply costs were on the high side was emphatic that they had to use higher cost supplies because they were getting better outcomes. Specifically, they insisted their complication rates were lower than their lower cost peers. Physicians are scientists by training and this was a testable hypothesis, so we pulled the data. We looked at the complication rate for Laparoscopic Cholecystectomy. In addition, this data was risk-adjusted and bench-marked to Premier's select data to reveal each surgeon's actual to observed ratio for complications. (See Graph 1)

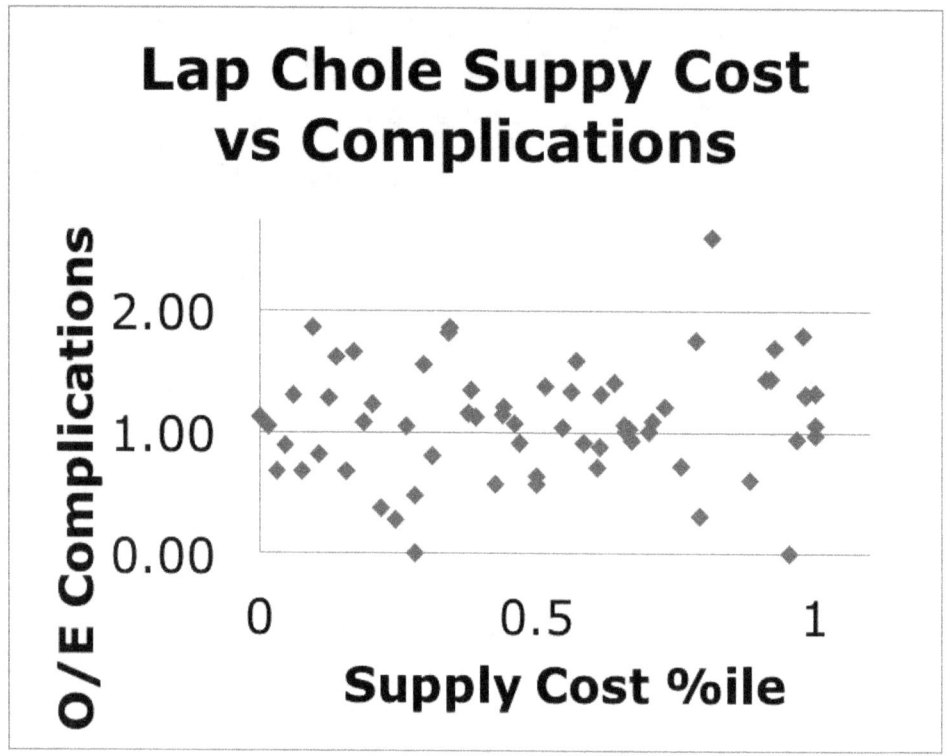

Graph 1: Comparison of average supply cost per surgeon to their observed to expected complications for laparoscopic cholecystectomy

Less than one is better than expected, and greater than one is worse than expected. We than compared this to the surgeon's average supply cost and ranked them by percentile from best to worst. There is no correlation between supply cost and complications. Needless to say, this generated some physician engagement, but it was not enough for most surgeons to want to run out and change their supply usage.

We then suggested that one of the ways they could reduce their supply cost is to start using reprocessed equipment. There was a great deal of heated discussion regarding reprocessed supplies and the consensus among the surgeons was that reprocessed (RP) supplies were defective more frequently than original equipment manufacturer (OEM) supplies. This too was a testable hypothesis. We tracked the surgeon reported defect rates for single use energy devices (not robotic). What we discovered was that the OEM defect rate was 4.9X greater than the RP defect rate[7]. Our reprocessing usage rate improved dramatically after this announcement as well as our supply cost savings.

> **Rule #4:** *If you want physician engagement, then appeal to their interest and passion for science.*

Informational asymmetry is generally considered a competitive advantage. This is where one company or team has access to information that the other company cannot access. It is an important mechanism in a competitive environment. Taken outside of this context it can have unintended consequences. For years' suppliers In the healthcare industry insisted on statements in their contracts with hospitals and healthcare systems which prevented them from discussing the prices of their products with their competitors.

---

[7] Loftus, T. A Comparison of the Defect Rate between Original Equipment Manufacturer and Reprocessed Single-use Bipolar and Ultrasound Diathermy Devices. J. Med. Devices. 2015; 9(4):044501-044501-2. doi:10.1115/1.4030858.

Understandable from a business perspective, however one of the consequences of these types of agreements is hospitals felt they were therefore prevented from having discussions with their physicians regarding the costs of supplies. This is simply not true. Hospitals can have these discussions with physicians. Unfortunately, this belief led to decades of physicians being prevented from understanding the cost of the supplies they used. Physicians have some understanding of the quality of the products they use. When the costs are hidden from them, it is almost impossible for them to understand the value of them.

While costs in healthcare are complicated, they are not too complicated to be understood by physicians. It just takes time and an interest to learn. We provided a group of General Surgeons with their average cost per case for Laparoscopic Cholecystectomy. This alone, was insufficient to impact any change in their average cost. In a blinded fashion we then showed them their average cost per case compared to their peers. Something in them changed. They were now very interested in cost. At one point, surgeons were now debating over who had the lowest cost per case. When they found out who it was, they wanted to see what it was that surgeon was doing. Their concern was how could this one surgeon possibly be beating them on average cost and still be getting good outcomes. For Laparoscopic Cholecystectomy most of the difference in cost comes down to a few instruments and several preference card practices, which we will discuss in a later chapter. Over the next several months the group reconciled the differences on their supply usage. They ultimately developed one single, low cost, preference card for the group. The group now sees themselves as the high value (high quality, low cost) alternative and markets themselves to payers as such.

> *Rule #5:* If you want physician engagement, then appeal to their competitive spirit.

This particular success story happened for another reason as well. The group was motivated by their competitive spirit. They not only wanted to be the best as individuals, but along the way they discovered a group identity that now wanted to also be know as the best group. Motivation only gets you part of the way. You can have all the motivation in the world, but if you do not have the ability to change, then change will not occur. So where did they get the ability to change. It came from themselves. They could see who among them had the low supply cost and best outcomes, and were able to model the group on the most cost-effective supply solution.

> **Rule #6:** *If you want physician engagement, then empower them to create their own solutions.*

Just as I was leaving work to go on vacation, I received a call from a very irate surgeon. He was upset and he needed to talk to someone from the administration. I was the Chief Medical Officer so I was used to these types of calls. This one was a little different. He spent the first half of our hour long conversation dropping just about every imaginable f-bomb you can hear. He was so loud and upset, I had to hold the phone away from my ear. During this time, I was occasionally scribbling notes on a note pad. The last half hour of our conversation was more amicable. We agreed to talk when I returned the following week. When I returned a week later there was a stack of letters, memos and reports I needed to review along with the other duties of my role. It was at least a week later when I found a yellow sticky note with a couple of ideas written on it. When I first read them, I thought, "these are some great ideas, why didn't' I think of them sooner". Just then it hit me. These were the scribbled notes I wrote when listening to the irate surgeon during the prior week. Buried beneath his anger, disappointment, negativity and multiple f-bombs were a couple of great ideas. Even though he was extremely upset, his ideas were what he was really trying to communicate. His passion was obscuring the real contribution. There are going to be times when you have to filter what you are experiencing when confronted with very passionate and angry staff. Beneath that passion you could very well discover some great ideas and engaged physicians.

> **Rule #7:** If you want physician engagement, then listen to them when they are upset, because they are already maximally engaged and may be harboring ideas worth considering.

The single most important thing you can do to engage people is to listen to them and understand what they are already engaged in thinking and doing. The next most important thing you can do is to appeal to what they value. The final thing you can do is to break down the silos that create asymmetric information. This can empower them to create their own solutions. Because the staff on the frontlines have first hand knowledge of the problems, they are more likely to implement a more cost-effective and sustainable solution.

One last word of advice regarding consistency. Good habits are the result of being consistent. By developing good habits on how you structure your program, and approach your stakeholders, you will appear consistent to them. By being consistent, you will build credibility and trust. People want to follow leaders they know they can trust, so, as a rule, if you want physician engagement, then be consistent.

CHAPTER 4:
COMMUNICATION

A thorough understanding of your stakeholders is the beginning of any effective communication plan. Leave one of them off the list of those with whom you will be communicating, and it could create some real problems for the success of your program. We're going to start with the most obvious group first, and that is the members of your Robotics Committee. They are not just members of your committee, they are also your program's ambassadors to the hospital and community. If this group does not understand, and cannot articulate what the program is doing, then other stakeholder groups will be receiving conflicting information.

Robotics Committee Communication

There are two basic documents used for communicating with your members. They are an agenda and meeting minutes. Do not neglect the importance of these documents. The agenda is a forward looking document. It describes what you will be addressing in the future. The meeting minutes are a backward looking document. It describes what we discussed and found important enough to put in writing. One looks at what we would like to accomplish as a team, and the other looks at what we were able to accomplish as a team.

There are two bad habits that committees can get into with these documents. The first is regarding the agenda. There is a tendency for committee leaders to issue the same agenda for every meeting. This sends the message that this next meeting will be no different than the last meeting. This is a disengaging message to your membership. The agenda must change if the program is to change. Ask your members for agenda ideas. Bring in a speaker to talk to your committee. There are going to be certain agenda items, like your dashboard, that will be recurring items. The point is to mix it up and make it interesting, even engaging, for your membership. How do you do that? Provide new information on the types of cases they are performing. This could also be a recent journal article, information from a meeting on new techniques, a video from YouTube demonstrating a novel approach to a common problem, etc. Make it interesting and they will want to be there.

The other bad habit has to do with the minutes. They are either too short and provide no useful information, or too long and detailed. Both types tend not to get read, although always approved. You want your minutes to reflect the discussion and especially any conclusions or action items for the team. You should be able to look, years from now, at any big decision the committee made back then, and understand why the committee made that decision. This is a very effective way to communicate to future leaders and members of your committee. Minutes aren't just for the current leaders and membership; they are for all groups who want to understand why things happened the way they did. What makes perfect sense today, but is not documented, may not appear obvious in a few years to those trying to understand your decisions.

Members should have the ability to contact one another on a regular basis. Create a membership list with contact information and distribute the list to all members (with their permission). There are going to be times when members will want to have an offline meeting to discuss some issue that they did not feel comfortable discussing in an open meeting. An offline meeting is a meeting that occurs between two or more members outside of the scheduled meeting. Minutes are not recorded for these types of meetings. These types of meetings are for members to clarify and work out any issues that were not, or are not going to be discussed in the open meeting.

One final note on Robotics Committee communication and that is where to store all official documents for the committee. Ideally this should be a cloud based repository (Dropbox, Google documents, etc.) or SharePoint. Committee members work all hours of the day and will need to be able to access this source of information. While it is important to email or otherwise distribute meeting materials, it is also essential for members to be able to review this information from anywhere and anytime.

Medical Staff Communication

Probably one of the most underappreciated teams in any hospital is the Medical Staff Services team. Much of what you read written in this book regarding infrastructure and communication was taught to me over the years by the people who ran the Medical Staff Services offices in the healthcare organizations where I worked. If your Robotics Committee needs to communicate to the Medical Staff, and by now you should know you will need to communicate to them, then you will need to bring the Medical Staff Service's office onboard early in the process. Regardless of whether you use a Medical Staff Model or an Administrative Model (more on this in the chapter on Accountability), it will be to your benefit to involve this team.

Medical Staff Services have mastered the art of organizing, communicating and coordinating the affairs of physicians and providers for decades. It has been referred to as "herding cats", but this is not accurate. It just seems like herding cats when you're not very good at it. Because of the skill set they bring to the table, this team can show you the way.

Here is one way they can do this. Early in the develop of our system Value Analysis Program, we needed a way to communicate to physicians. I contacted our Director's Group for Medical Staff Services and requested the email contact information for all of our Medical Staff. They refused. 'Why", I asked. "Because that is confidential information", they responded. Good point, I forgot about that. They also pointed out that the list changes every day and it would be impossible for me to update it all the time. Good point, again. They reminded me that some of the physicians do not check their email or spam all the email coming from the hospital. Didn't know that, but also another good point. What do they do in that situation? It depends. Sometimes they send an email to the physician's secretary to print it out and hand it to them. Sometimes they catch the physician in the lunch room or rounding on patients and talk to them. The bottom line is they have learned how to contact every physician on staff, when they really need to contact them. We then worked out an arrangement where I would forward the Directors of Medical Staff Services a newsletter that summarized the committee's finding. They would then forward it to the appropriate physicians and Medical Staff meetings. Sometimes it doesn't always work as planned, but it is safe to say, it was by far, one of the most consistent ways I found to communicate to large groups of physicians.

While this method is consistent, the real question is, was it effective? Physicians are busy people and let's face it their time comes at a premium. When you have their attention, maximize the use of that time. Wasting it, is another form of disengaging behavior. At the end of this chapter is an example of a newsletter we did for supply chain services. Note a few things. All the monthly news fits on one page. Each section is clearly marked so a physician can look at this with one glance, and in a few seconds see if there is anything of interest to them. The most important words and sentences are bolded so a physician can quickly read the highlights of the section to see if the information applies to their practice. If there is more information, then a memo is provided along with the newsletter and is available to the physicians who want to read additional information. There are links in the newsletter so if a physician is interested in additional information, then they can simply click on the link. This is when a SharePoint is especially useful because you can store and link the document in one place. This newsletter also includes an email link to the leadership in case a physician has any questions or comments. Send a few of these out and notice the response. You will quickly find out if you are being effective or not. Physicians tend not to hold back when they really want to get a point across.

If only newsletters and memos were all it took to communicate to your Medical Staff. Nothing does it better than a face-to-face conversation. There are going to be times when you must get up in front of a group of your peers and explain why you are doing what you are doing. They may not agree with you, but if you are not willing to face them, then they will lose trust in you. When it comes to the relationships you have with your Medical Staff, trust is the cornerstone. The CEO of a company where I worked used to remind us of an old saying. Visibility leads to credibility and credibility leads to trust. If you want to be trusted, then you must be visible. If your Medical Staff are complaining about communication, then it is time to get visible.

C-Suite Communication

If the Administrative Lead on your committee is a member of the C-Suite, then this will be very easy. If they are not, then you will need to make sure that what is happening on your committee is being communicated to them. The C-Suite is made up of the senior management team for your facility. They will be making all of the major decisions regarding operations and capital purchasing. If you have a robot at your facility, then this team authorized the purchase. They understand this is an investment and will usually be very engaged in learning about how that investment is progressing. While they will enjoy hearing about what a great job you are doing to improve clinical outcomes, they will be especially interested to hear about the improvements in utilization and cost reduction. They, after all, want to know what the return on investment is. This is also a great opportunity to let them know what resources you need. Your Administrative Lead should be running point on this. If you are not able to improve your performance because you lack adequate resources, then this team needs to be aware of this situation. One last word of advice, whatever you are communicating to your Medical Staff, it must me the same to your C-Suite. Believe it or not, they talk to each other. Well, at least in most organizations. So don't forget to include your C-Suite on the Newsletter or any memo distribution list.

Nursing & Ancillary Staff Communication

This tends to be easier to do than physician communication. Just because it may be easier does not mean it can be neglected. Typically, the Chief Nursing Officer and this person's direct reports are your best avenue for communicating. Coordinating with this person early and regularly is important. You can use a newsletter or any other approved facility communication tool. Be open to feedback, particularly when it comes from the operating room. If you do not have your Robotics Coordinator, OR Director or Nurse Manager over robotics on your committee, then this is a good time to include them. This person will know the key players and how other operations will impact your program.

Community Communication

If you haven't done it already, then it is a good time to bring your Marketing and Public Relations team on board. You will want to make sure your community is aware of your ability to provide this service. The internet and your website is a great place to start. More and more people use the internet to research their options when it comes to healthcare. Patients who use the internet tend to be more informed and also tend to have commercial insurance. If they are interested in a robotic assisted procedure and they are not aware you offer it, then they are most likely going to take their business elsewhere. Like it or not, it is just the way it is these days.

Supply Chain Service's Newsletter

What's New in November, 2014

Orthobiologics: DBM

- Earlier this year an Orthobiologics **Value Analysis Team composed of Surgeons** met to discuss the benefits and cost-effectiveness of multiple products in this supply category. Based on the **recommendation** of this team, effective immediately, **The system will no longer purchase or contract for the following Demineralized Bone Matrix (DBM) products:**
    - Stryker AlloFuse Gel, AlloFuse Putty
    - Stryker AlloFuse Plus Paste, AlloFuse Plus Putty
    - Stryker AlloMatrix Putty
    - Stryker Allograft Wedges
    - Stryker DBM Gel, Putty & Putty Plus
    - Integra Dynagraft products
- **All other contracts for DBM products remain in effect.** We do have existing contracts with Bacterin, Medtronic, MTF, and others for DBM products and will continue to support this product category.
- **See attached memo.**

Irrigation Fluid Shortage

- **Conservation methods are working.** At our current rate of use we are on track to meet the usage needs into the foreseeable future. Since this shortage is anticipated to last into the first quarter of 2015, **we are asking everyone to continue to conserve irrigation fluid.** The effort on the part of the physicians, staff and Supply Chain has made all the difference. **Thank you for your assistance!**

Supply Chain Services Link

Additional Information
- Click on this link

Supply Savings Strategic Initiative

- The **Overall Supply Savings Strategic Initiative** results, as of Nov 3, 2014, are in. **We exceeded our stretch target** for the system and are at $28,632,916 for the year. Peri-op is 106% of the way at $7.54M and Cath Lab is 126% of the way at $6.09 M to their target. **Congratulations** to the facility and discipline teams for their outstanding performance!

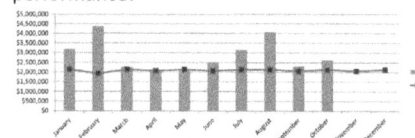

Supply Chain Education & CME

The **American Association for Physician Leadership** (formerly ACPE) is offering a course titled, **"Physician Leader's Role in Supply Chain Performance"** (CME = 14hrs) at the Fall Institute in Scottsdale, AZ (Fairmont Scottsdale Princess) on Nov 16–17. Contact AAPL for more info.

Safe Surgery

The results of the system's Safe Surgery Program were published online in the **Journal of the American College of Surgeons.** The title is **"Implementing a Standardized Safe Surgery Program Reduces Serious Reportable Events".** There was a **52% decrease in the SRE rate.**

Questions or Feedback

Contact: Terry Loftus, MD, FACS
TerryLoftusMD@gmail.com
Medical Director
Surgical Services & Clinical Resources

CHAPTER 5
INFRASTRUCTURE

Infrastructure is the foundation on which any program will be built. There are elements of infrastructure which will be foundational for all types of programs and elements that are specific to a Robotics Program. We will cover both in this chapter. All successful programs have a well developed infrastructure. Unsuccessful programs and poorly performing programs, all too commonly, can be found to be lacking in basic infrastructure. So what exactly do we mean when we say infrastructure.

INFRASTRUCTURE DEFINED

Infrastructure refers to all the people, process, technology and cultural elements that are required for the successful implementation, management and sustainability of a program. The specific elements are categorized according to people, process, technology and culture.

PEOPLE

People are the personnel who are essential for the smooth operations of a Robotics Program. They are characterized by the roles and responsibilities they have as members of the Robotics Program. The following roles and responsibilities are considered fundamental to any successful program.

1) **Physician Lead:** This individual provides direct oversight of the Robotics Program. This role can be developed as a sole leadership role or supported by a co-lead in a dyad model. Most commonly a dyad model in a hospital setting would be with a Nursing lead. Some systems may choose an Administrative lead as a dyad co-lead. This leadership role functions as the Chair of the Robotics Program and Robotics Committee for the Hospital or Healthcare System. They are responsible for setting the goals of the program, developing an agenda for meetings, running the meetings, developing the program/meeting membership, proposing the purpose of the program, develops goals for the program and is the primary person responsible for communicating any information regarding the program to the Medical Staff and leadership. In addition, this person is also the individual responsible for receiving and responding to feedback from the Medical Staff and leadership. This person should be a surgeon who is active in the practice of robotic assisted surgery. The Physician lead, along with the Nursing lead, provides oversight for the Robotics Program's performance improvement.

2) **Nursing Lead:** This individual, when functioning as a dyad co-lead, has the same responsibilities as the Physician lead. The Nursing lead is the person responsible for communicating any information regarding the program to the Nursing and Ancillary staff. In addition, this person is also the individual responsible for receiving and responding to feedback from the Nursing and Ancillary Staff. This person should be a nurse who is either active in the practice of robotic assisted surgery or provides immediate oversight to those areas (e.g. Robotics Coordinator). The Nursing lead, along with the Physician lead provides oversight for the Robotics Program's performance improvement.

3) **Administrative Lead:** This individual can be paired with a Physician lead in a dyad model to co-lead the program. If so, then the Administrative lead's responsibilities are similar to the Nursing co-lead's responsibilities. The Administrative lead's primary function is to be the liaison between the hospital administration and the Robotics Program. This individual is responsible for communicating any information regarding the program to all of the staff (Non-Physician/Non-Nursing) and especially the C-Suite. This person should be a senior level manager who has direct access to the C-Suite, and if possible is a member of the C-Suite. Ideally this person should have operational and management oversight of the areas impacted by the use of the robot. One of the main functions of the Administrative lead is to identify and secure resources necessary for the smooth functioning and sustainability of the program.

4) **Anesthesia Lead:** Depending on the needs of the program this person can also function as the Physician Lead. More commonly the Physician Lead role will be filled by a surgeon who is active in the practice of robotic assisted surgery. When not in the Physician Lead role, the Anesthesia Lead is an Anesthesiologist or Certified Registered Nurse Anesthetist (CRNA) who is active in clinical practice and provides anesthetic services to patients undergoing robotic assisted surgery. The Anesthesia Lead's primary function is to be the liaison between the Medical Staff's providers of anesthesia services and the Robotics Program. In addition, this person will provide guidance and support for the program's performance improvement.

5) **Administrative Assistant:** This person provides administrative and clerical support for the program. Do not, repeat DO NOT, underestimate the importance of this role. This is a critical role for the success of your program. Duties for this person include: setting up meetings, finding meeting rooms, distributing meeting notices and agendas, records meeting minutes, stores meeting documents, maintains member contact list and coordinates

RSVPs. This is just a partial list. Hopefully you get the point. This person does just about everything to make sure the program functions efficiently and effectively. This role requires a professional assistant with experience organizing meetings, especially meetings with Medical Staff.

6) **Data Analyst:** A highly functioning Robotics Program must have access to data and someone trained in data analysis. The data must be consistent and whenever possible risk-adjusted. Bench-marked data is also very useful even when a program is using itself as a bench-mark. Bench-marked data to other healthcare systems can also be very useful. In the beginning, a program must establish a baseline and compare this baseline to itself over time and to others when such comparisons are available. If the goal is performance improvement, the program must know where it stands in order to understand where it needs to go.

7) **Multidisciplinary Representation:** The surgeons who use the robot need to be represented on any committee that is part of your Robotics Program. It is not necessary to have everyone who uses the robot on the Robotics Committee, but every specialty that uses it should be represented.

PROCESS

1) **Regular meetings:** If you want your Robotics Program to succeed, you will need to have regular Robotics Committee meetings. How often is up to the program's leadership. As a general guide, the busier your program, the more frequently you will need to meet. If you have one robot and do less than 240 cases/year, then once a quarter is probably sufficient. If you have 3 robots and are doing 1000 cases/year then you may need to meet on a monthly basis. One program we are aware of has 5 robots and performs over 1500 cases/year. They meet

weekly and the CEO of the hospital attends each meeting. Always schedule your meetings far in advance and on a specific day, time and location. (e.g. first Monday of the month at 5:00 PM in Conference Room A) As we will learn in the chapter on engagement, consistency is important for driving engagement.

2) **Maintain an Agenda and Meeting Minutes:** The agenda is used to drive the discussion and the meeting minutes will document what was discussed. The agenda needs to be shared, in advance, with the Robotics Committee membership. Members must play a role in shaping the agenda. The best way to do this it to list an agenda item, usually toward the end of the meeting, of ideas for future agenda items. Another way to do this it to request agenda items be forwarded to your administrative assistant or co-leads. This communicates to your members that you are seeking their input on how the program will be managed.

3) **Established Reporting Structure:** The Robotics Program should report to the hospital's Medical Staff and Administrative leadership. The model you choose is up to the individual program. There are two basic models of governance. One is Medical Staff and the other is Administrative. A Medical Staff Model is one where the Robotics Committee reports up through a department or directly to the Medical Executive Committee (MEC). In this model a physician is the sole lead and non-physicians are invited guests. In an Administrative Model, the Robotics Committee reports up through an Administrative department or directly to the C-Suite. Leadership based on a dyad model is possible with this approach. Non-physicians are typically equal members of the committee. Regardless of which model you choose; the important feature is that the Robotics Program is reporting up to leadership. It creates the governance for your committee and provides credibility for it as well. Highly functioning programs will typically report to both the MEC and the C-Suite. This allows for better communication between

Medical Staff and the Administration. We will discuss this more in the chapter on Accountability.

4) **Communication Plan:** The Robotics Program must have a consistent method of communicating important information from the program to its stakeholders, as well as a consistent method of hearing back from them. This can be everything from a monthly newsletter, to a blast email, to regular town hall meetings. It is entirely dependent on the size, complexity and culture of your hospital. Either way it must be regular, consistent, easy to access, easy to interpret and reliable. It typically will come directly from the Robotics Committee or its leadership. Most programs will have various formats for communication based on the information being communicated and the intended audience. Performance improvement information will typically be communicated to the clinical staff and C-Suite, whereas education may have a much more selected audience, e.g. scrub techs assisting with robotic assisted surgery.

5) **Data Review:** One of the first items a Robotics Program will need to determine is what metrics it will track to determine if its performance improvement plan is working or not. In the chapter on performance improvement, we will provide a list of metrics. One of the biggest problems a program may encounter is to have data being generated at some level in the organization, and not have a process for accessing the data related to its programs. If your system has an enterprise data warehouse (EDW) this process may already be established. Healthcare systems, like most large companies, will often have multiple data systems with multiple operators managing those systems. An EDW pulls data from all or many of those systems into one data warehouse. This allows for better data management and the ability to analyze data from multiple sources at once. Systems with an EDW have usually also created processes for pulling this data for analysis. If your system is set up like this,

then you will need to tie into this process. If not, then you will inevitably be working with several people who will need to be coordinated for data abstraction and analysis. Focus first on the data you can get. Things like number of cases, turnover times and robotic room utilization are usually tracked by your perioperative team. If so, then don't forget to invite them to your meetings. If you are getting push-back from the people who control the data flow, then it is time to work with your Administrative Lead to find a way to get the data. Once you have determined the data you want and who controls access to it, decide how often your committee needs to see the data. Data must lead to some action so it is important to have enough data to feel comfortable acting on it. If you program does 240 cases/year, then this is 20 cases/month. Monthly reporting is going to be too frequent. Quarterly or every 6 months may be more appropriate. Some data (mortality rate) may need a less frequent reporting period (annual). To summarize:
   a. Determine the data to be tracked
   b. Develop a process for abstracting and analyzing the data.
   c. Determine the frequency of data reporting for all data points.
   d. Create a dashboard for data reporting.
6) **Goal Setting:** This will need to be coordinated with #5 above. Have some idea of what goals you will have for your program. For example, if your goal is to reduce close to cut time, then close to cut time will need to be included on your dashboard. (This is the time from when the incision is closed in one case to the time when the incision is made on the next patient. Believe it or not, this time is correlated with surgeon satisfaction.) In addition, you will need to think about component goals. For close to cut time, you may also find it important to track component times of this time, such as close to out of OR time, Out of OR time to In to OR time, and In to OR time to cut time. The reason for this is, any performance improvement efforts can

be focused on the component time that is an outlier. If the close to Out of OR time is much greater than expected, then the issue may be with how Anesthesia is waking the patient at the end of the case. In addition, goal setting should be realistic. Making goals such as we will be in the top 10th percentile in every metric in the next 3 months is not realistic. A goal such as, we will reduce OR turnover time from 32 minutes down to 27 minutes in the next 6 months is more realistic. If you have never used the S.M.A.R.T. approach to goal setting, then now is the time. S.M.A.R.T. stands for specific, measurable, achievable, results-focused and time-bound. (See above example.) Use this or a similar method of setting goals for all of your program's goals.

7) **Credentialing and Privileges:** Credentialing and Privileges is generally considered a Medical Staff duty. Regardless of whether your Robotics program chooses a Medical Staff Model or Administrative Model, the Medical Staff who are members of your program should be considered subject matter experts for robotically assisted surgery. How this group participates in this process for the Medical Staff office is an essential function of the group. There is no one better qualified to perform this function for your organization. Integrating this group into that process should be a priority for the program.

TECHNOLOGY

1) **Robot/Case Mix Alignment:** Challenges in the operating room environment led to advancements in robotic technology. This created the need for differentiation among the robotic platforms. Having the appropriate platform for the types of cases a system is performing, demonstrates a more strategic approach to program development. If your program has a high volume of colorectal and ventral hernia repair cases, and your surgeons are working with older technology, then this can create some degree of frustration among the staff. It could also be impacting your utilization. Robotics programs should strive to have the best mix

of robotic technology and case mix. If your program has a relatively high volume of cases, then having a portfolio of robotic platforms may be a more cost-effective way to manage the program.

2) **Electronic Medical Record:** There are two important reasons why an EMR is good to have in a Robotics Program. When you have access to an EMR, this creates opportunities to standardize order sets and collect data. Both of these are key to a successful quality improvement program. Standardized order sets allow your program to create care pathways which are associated with improved outcomes. The data generated from your EMR allows your program to monitor whether those care pathways are working, and if they are producing the clinical, utilization and cost improvement goals your program established.

3) **Risk-adjustment and Bench-marking Capability:** There are a number of valid programs available in the market place and many large systems have access to this capability. What you will need to consider is whether they are based on administrative data or clinical data. Each has its pros and cons. If you have access to both types, then consider your program fortunate. Administrative data is based on coding, which is dependent on documentation. Healthcare systems with strong documentation programs tend to shine in these types of databases so your external bench marks may be somewhat deceiving compared to these organizations. Do not despair, they can still be useful. First, you can always use them to compare your current data to your historical data. When you implement a quality improvement program you can still track your progress in a before and after manner. Clinical databases have some limitations as well. They typically cost more to manage and therefore many programs will only abstract random cases to enter into the database. If you have a low volume program, then it will take much longer to generate statistically significant

data. The information you get will provide a more meaningful report of your clinical performance.

4) **Centralized Storage for Documents:** All documents associated with your Robotics Program will need to be stored electronically in a common safe place. Typically, this can be with a SharePoint or some other type of cloud based storage. Members must have access to this repository. This is where you will place documents such as meeting agendas, meeting minutes, dashboards, literature, policy statements regarding robotics, block time, etc. Avoid placing this on someone's (Administrative Assistant) computer. If that person leaves the company, there is the risk that everything you've done for the program could be lost as well.

5) **Tele-conferencing Capability:** There are going to be times when members of your program will need to participate from a remote location. At the very least, programs should have a call-in number for members who can't physically be present. Make sure your rules allow for members to call in remotely, and have that count as a quorum. It is also important to have a consistent call-in number and not change it meeting to meeting. You are more likely to get participation when the call-in number and any live-meeting access is consistent. (See chapter on Engagement.)

CULTURE

1) **Quality Improvement Program:** All hospitals should have a quality improvement program. The question is, how active is your program. If the only projects your quality department works on are required for legal, compliance or regulatory purposes, then it falls in the category of just above the minimal threshold. Quality departments that are actively working on other projects, in addition to the minimum, are demonstrating a cultural attribute consistent with high performing organizations. If your Quality department is actively working on a Robotics

quality improvement project, then you are well on the way to success.

2) **Epicenter/Center of Excellence:** These are typically designations that can be from external agencies or simply one developed by the hospital marketing department. It is important to differentiate what the motivation for the designation was. In a competitive healthcare environment, an aggressive marketing program is understandable. If volume is the only metric that is improving, then you probably have a pure play marketing reason for this designation. While volume can be associated with quality, this is not always the case. When quality and utilization improvement accompany this designation, then you are well on the way to success.

3) **Quality Awards:** Has your hospital or healthcare system received any awards for performance? This may include Truven Top 100 Hospital, Truven Top 15 Healthcare System, Baldridge Performance Excellence Award or any similar type of recognition for performance improvement. If so, then there is a good chance you are part of a culture that is going to be very interested in seeing your Robotics Program develop. While these awards are not specific to Robotics, they are markers of a culture that understands the importance of performance improvement.

4) *Robotics Supportive Environment:* Do the members of your program feel supported? Is there dedicated block time for access to the robot? Does your administration listen and respond to your surgeon's needs regarding equipment and the logistical support for your program? Do your departments and MEC listen to and respond to your surgeon's issues related to robotics. If the answer to any of these questions is "no", then your local culture will become a major challenge to the sustainability of your program.

CHAPTER 6
ACCOUNTABILITY

One of the first decisions you will need to make is what type of accountability model your Robotics Program will use. This will determine what type of governance your program will have. There are two basic models currently in practice in facilities that have a program. Which one to use is entirely up to the individuals who are charged with developing a program. The two models are the Medical Staff Model and the Administrative Model. We will discuss the pros and cons of each model, which you can use in your decision.

The Medical Staff Model

The Medical Staff Model is where the Robotics Program is established and reports directly to either a division, department or Medical Executive Committee (MEC) of the Medical Staff. Because it is a formal part of the Medical Staff structure it is accountable to and responsible for any Medical Staff bylaws or rules and regulations.

The biggest advantage of a Medical Staff Model is that your program will have an established infrastructure. The stronger your Medical Staff infrastructure and the greater the support you have from your

Medical Staff Services department, the greater the likelihood you will have an effective program compared to not having an organized structure. Medical Staff Services will typically provide the logistical support your program will need. They will set up meetings, send out meeting documents, notify members of meetings, help set agendas, record meeting minutes, and store all meeting documents. They will help establish the foundation for your program, because that is one of their fundamental functions, and as a general rule they are usually very good at this job.

When a Robotics Committee is established within a Medical Staff Model, it is usually established as a division or subcommittee under an established division or department. This means a Chair will be named by the committee to whom it will be reporting. Membership will consist of members of the Medical Staff and only these members will have voting rights. Voting will typically be done, only when a quorum is established by whatever rules your Medical Staff has regarding voting and what constitutes a quorum. While there can be a Vice-Chair, this person would also need to be a member of the Medical Staff. Any other members who are not Medical Staff (Nursing, Quality, Data Analyst, Business Analyst, Administration) are generally considered invited guests.

A Robotics Program within a Medical Staff Model also has an established accountability structure. The Robotics Committee reports to and is accountable to the committee that established it. This committee will ultimately report to the MEC. The MEC will be comprised of Medical Staff leadership and Administrative leadership. While the administration will not typically have voting rights, they will be able to provide this group access to information regarding operations and financial performance. In addition, the MEC will have the ear of the administration, and if there are specific resource issues related to your program, support from both the MEC and administration will go a long way toward securing those resources.

Another advantage to the Medical Staff Model is its ability to leverage peer review to improve quality at the surgeon level. Peer review would either occur at the committee, department or facility level depending on how your peer review is structured. This provides legal protection, based on state law, for executive sessions in which physician level discussions occur for cases identified for peer review. Because this is occurring with in a Medical Staff Model, these discussions will need to occur within the rules for your process. With the exception of staff from your quality office and select administrators, the invited guests will not be able to be present during these peer review meetings. With all of these advantages of a Medical Staff Model, why would any facility not want to choose this model for its Robotics Program.

One of the first questions you will need to answer with a Medical Staff Model is which department or division will it be reporting? If you have only one specialty using the robot, then this will be very easy to answer. If you have more than one specialty, then this will require much more discussion. In one organization, they started the program with just one specialty but as additional specialties began using the robot, they were often left out of the committee discussion. When they were present at the committee meetings, they were treated as invited guests and did not have voting rights. In this case, the OB/Gyn Department created a Robotics Committee when only gynecologists were using the robot. When the facility recruited urologists and general surgeons who would use the robot, problems developed in terms of access to the robot and block time. What happened was a highly functioning committee became dysfunctional for the organization. Exclusivity is incompatible for the success of your program when it becomes a multidisciplinary program, but lacks fair representation at the committee level.

Another problem is the Medical Staff Model can not only exclude other members of the Medical Staff but could easily start excluding participation and support from Nursing, Administration and Ancillary

Staff. Disregarding these key stakeholder groups in your program could create some real problems in terms of outcomes. If the program encounters capacity issues, it will be essential to have them at the table early. If you want to expand from eight to twelve hour days or open the robot for weekend use, then you will be running up against every administrator's nightmare of premium labor. Premium labor is when the hospital must pay staff more for being on duty. This includes shift differential pay, weekend/holiday pay, overtime pay and paying registry to cover the additional shifts. Committing to additional capacity can very quickly double the costs of this increased capacity, if not approached in a very strategic manner. In addition, the cost to train this workforce on robotic assisted surgery or find registry personnel already trained is not inconsequential. The growth and support of your program is going to be dependent on these stakeholders. Be sure to include them in your program from the very start.

Administrative Model

With all of the potential advantages of a Medical Staff Model, why would you choose an Administrative Model? This is a model that provides a great deal of flexibility and some healthcare organizations are going to need to be flexible, particularly if your Medical Staff does not think they need another committee or your Medical Staff committees are not very effective at improving quality, utilization and cost. In addition, they may not be willing to create a multidisciplinary committee that reports only to the MEC. If you are having any trouble developing a Robotics Program due to lack of support on the part of the Medical Staff leadership, then this may be a solution.

The Administrative Model is supported with the assistance of your organization's administration. The administration will provide the resources (Administrative Assistant, Data Analyst, Quality Department support, meeting rooms, etc.). The downside of this is the administrative support for your infrastructure may not be very experienced at organizing meetings with physicians. Medical Staff Services has typically reduced or eliminated those disengaging behaviors from years of experience working with physicians. People new to this experience will need to learn this very quickly or your program may suffer the consequences. One solution to this problem is to recruit your Medical Staff Services office to do this for your committee. This may prove to be difficult. If your Medical Staff is not willing to support a new committee within its existing structure, then they tend to be somewhat hostile to the idea of lending out their support services to perform this function to the hospital. It can't hurt to ask, but remember to have a back-up plan if the answer is "no".

Leadership for this committee is typically set up as a dyad. Usually this is one Physician Lead and a Nursing or Administrative Lead. It will report directly or indirectly to the C-Suite. The best of these models has a matrix reporting structure to the Medical Staff. This can be where the Robotics Committee's Physician Lead is either a member of a Department Committee or MEC. Another arrangement may be where the Physician Lead or Nursing Lead is an invited quest at these committees and reports on the progress of the committee. The reason this is an optimal approach for an Administrative Model is, it maximizes the communication to both the hospital administration and the Medical Staff.

One of the major advantages this model supports is it can truly be a multidisciplinary committee made up of different surgical specialties (OB/Gyn, Urology, General Surgery, ENT, Cardiothoracic, Anesthesia, etc.), different nursing departments (Perioperative, ICU, Med-Surg, etc.) and different administrative departments (Quality, Data, Clinical Informatics, Finance, Supply Chain, Marketing, etc.). This increases the information available to the committee and allows for greater stakeholder support. The trick is coordinating these various stakeholder groups. More on this in the chapter on Leadership. In some ways, robotics is forcing hospitals to consider these types of structures in order to maximize the performance of their investments.

A similar thing was observed with hybrid vascular labs. In one facility, the administration was reluctant to invest millions of dollars in a hybrid operating room until the Cardiothoracic Surgeons, Vascular Surgeons, Interventional Cardiologists and Interventional Radiologists were able to meet first, and discuss how this room would be utilized for the benefit of patient care. This led to a multidisciplinary committee and the successful launch of a highly productive and cost-effective hybrid OR.

There will inevitably be cross disciplinary debates about scheduling, block time, appropriate use, training, supplies, etc. It will be to your advantage to have a multidisciplinary committee to bring these issues. If there are issues that directly impact the work of Medical Staff (peer review, credentialing, privileges, appointment and reappointment criteria), then this group can serve an advisory function to the appropriate Medical Staff Committee, or in the case of peer review forward any information to the appropriate department for formal review.

Alternative Model

While the Medical Staff or Administrative Models are the most common models available, it would be incomplete to say they are the only models available. There are hybrid models that capitalize on the best of both models. That is, use the Medical Staff Model's infrastructure and the Administrative Models multidisciplinary approach to optimize its effectiveness. It takes some creativity and quite a bit of cooperation from both the Medical Staff and Administration to accomplish this. If you have a culture of doing this, then more power to you. If you are doing this at a system level, then this type of governance could provide direct reporting up to a service line or another system level division that directly reports to the senior management of the organization. Another model we have seen used is, instead of calling it the Robotics Committee, it becomes the Minimally Invasive Surgery Committee, Division or Department for the hospital. Robotics is a subsection within this group. This is more commonly found at very busy hospitals that are promoting their Minimally Invasive focus for investigative or marketing purposes.

So which model is the best one for your program? The best one is the one that works the best for your organization. All of them have the potential to work. The ones that don't work very well are the ones that lack one or more of the seven pillars of successful programs. There are many ways to do it wrong, and only few ways to do it right. Do more of what you know works, and less of what you now know doesn't work.

CHAPTER 7
LEADERSHIP

The CEO of a healthcare system were I worked, constantly reminded us that leadership matters. When it comes to developing your Robotics Program, or for that matter any program, this is going to be one of the most important things for you to remember. If you have read this far into the book, then guess what? There is a very good chance you are either a leader, or an aspiring leader. Implementing and sustaining your program will be entirely dependent on your leadership. Yes, every chapter you have read so far is very important, but without leadership, the rest of the seven pillars will not get you very far. In this chapter we will discuss some of the key takeaways regarding the important behaviors of successful leaders and how to use leadership to maximize the effectiveness of your Robotics Program.

How can something so important as leadership seem so elusive? Part of the reason is because leadership means different things to different people. While there are certain qualities that most people will want to see in their leader (courage, integrity, calm under pressure, etc.), there are expectations of a leader that can vary significantly depending on the stakeholder group to whom they may be interacting. Awareness of those expectations and understanding why they are important is half the battle. We'll look at each of the major stakeholder groups with a particular focus on what they will expect of you as a leader and why.

Physicians

A few years ago I was addressing our Chief Medical Officer (CMO) meeting about a particular supply chain issue. I was looking for their support and assumed I had it before I walked into the room. It became obvious in just a few short minutes after I began the presentation that I was not going to get their support. Afterward a colleague approached me and gave me some advice. He said it sounded like I was speaking to them as CMOs, and as part of their facility's administration, which was true. What I forgot was that all of them are physicians first, and I did not speak to them from this perspective. He was right. Physicians, regardless of title, will always need to be addressed as physicians first. So what do I mean by this?

Physicians value what is right for patient care and the scientific method. This is how they were trained and it is what they know and understand. Approaching them with the business case for using a particular supply over another is probably not going to win you many friends with a physician audience. You may be able to make this case to the hospital administration, but, as a general rule, it will fall on deaf ears with a physician. They need to know from the beginning that quality and patient safety is the overriding concern. They understand the importance of improving utilization and reducing cost for the hospital, they just need assurance that utilization and cost improvements are not being achieved by subordinating quality and safety for patients and their community.

There is also something else that is very near and dear to physicians. As a group, physicians value autonomy and independence. In particular, they want to be able to practice without the sense that everything they do needs to be micromanaged by some outside entity. It is not that they do not understand the importance of guidelines, best practice and the merits of standardization - they do. They want a say in how they will be expected to practice and clear expectations on performance. Just as words like "compliance", "policy" and "mandates" tend to be disengaging to a physician, leaders who do not listen to what physicians are saying, and do not articulate performance expectations will, themselves, become disengaging to physicians.

Leaders who are attempting to appease one stakeholder group (hospital administration) while addressing another stakeholder group (physicians) will quickly find themselves fighting a losing battle. The last thing you, as a leader, need is to be perceived as a puppet of the administration. It is okay to focus your team on quality outcomes. If you can do this, then the road to improving utilization and cost will be much easier. This is for two reasons. The first reason is because if your physician group knows you are doing what you are doing for patient care, then they are more willing to work with you on other issues such as utilization and cost. The other reason is because excellent quality inevitably leads to better utilization and total cost of care savings.

Administration

Not only will different stakeholder groups have varying expectations of you as a leader, but don't be surprised if individuals within those stakeholder groups also have varying expectations of you. Early in a new role as a physician executive, I met with multiple leaders across the various regions of our organization. I had two questions for each of them. What are your immediate expectations of me and how would you define success for me in this role, one year from now? All of them gave the same vague responses such as, "more physician engagement", show "courage", "influence physicians" and my favorite, "get doctors to do more of what we want them to do". Needless to say these are not only vague responses, but also unrealistic because they are not well defined. I pressed them further for specific metrics or at least some type of semi-quantified definitions. This is when it got interesting. Each person had a different answer. They were using the same words but meant something different when they used it. I wish I could say this problem was confined to administrative leaders. It is not. I found the same problem with physician leaders. There is a solution for this problem that is not dependent on the other person.

Define success for them. When an administrator says they want to see more engagement from physicians as a result of your leadership on the Robotics Committee, then tell them what this means to you and get them to agree to that definition. For example, you can say that your first year goals for the Robotics Committee are to define a quality, utilization and cost savings goal for the team and realize a ten percent improvement in each. Then ask the administrator if that would qualify as more engagement from physicians? It is difficult for them to not agree with this because there is a very good chance that no one else has done this or suggested this. It is important for you to think about these promises before you enter into these types of discussions. You want to set achievable goals for yourself. So what makes this an attractive offer to an administrator? It appeals to something they value.

The administration is charged with running the operations of the hospital or healthcare system. They cannot ignore business operations. It is their job, and it is what they are trained to do. Their leadership will be holding them accountable to this. Any discussion you have with them must take this into consideration. In the example above, you will notice that goals would be set for quality, utilization and cost savings. It doesn't say what order these goals will be set. When you meet with your committee you can remind them that you will be setting these three types of goals, but you can focus first on quality. If your surgeons are most concerned with turnover time, then you can use that as your first utilization goal. The idea is to introduce goals as a way for the various stakeholder groups to achieve outcomes that are important to them and the organization.

Your administration may have specific goals for you to achieve. In this case it will be important to prioritize these goals and set realistic expectations. Use this as an opportunity to shape what their expectation of success for your role will look like. Programs take time to build and great outcomes will take even longer. If you are being asked to triple volumes, cut OR time in half and reduce supply cost by eighty percent, then this is a time to speak up and set realistic expectations. Most of you will encounter reasonable administrators who will work with you. They expect you to take the lead on this, so be brave and just do it. It helps to network with other Robotics Committee leads and have some understanding of the types of goals you can accomplish in the first year. It may be as simple as establishing a database to track your performance. That is a common problem and therefore a very realistic goal.

Nursing

Regardless of whether you choose a Medical Staff Model or an Administrative Model, you will most likely include a Robotics Coordinator or perioperative equivalent at your meetings. This individual will either be your Nursing Lead or work closely with your Nursing Lead and be the primary liaison with the nursing staff. The expectations of the Nursing Staff will lie somewhere between that of the physicians and the administration. While they are trained from a clinical perspective to put the patient's interest first, they will also be employees of the hospital. If they are in management, then there is a good chance that person will be responsible for a budget and may have utilization and cost savings goals for their department. There is nothing wrong with this, but you will need to be mindful of this when choosing your annual goals. The closer you are aligned with their goals, and what they are being incentivized to achieve, the greater the likelihood for your success.

There is another aspect of leadership that will be expected of you. This has to do with conflict resolution. It is not that the other stakeholder groups will not expect this of you, it is just that the group that will expect it the most will be Nursing. This can happen in many ways but the most frequent way is when there is a conflict with a physician. As a surgeon it breaks my heart to say this, but it will most commonly be with a surgeon. It's easy to pass this off to the Department Chair, Chief of Staff or Chief Medical Officer, but all eyes will be on you. If you are a physician or nursing lead over robotics and there is a conflict in this area, you will be expected, at the very least, to be aware of it and in many cases to directly intervene. There may be a tendency to shy away from this role but you really should not for three reasons. First, all eyes are watching you and want to see how you respond as a leader. Second, if you can handle this well it is a great opportunity to gain the respect of your team. Finally, as a Leader of Robotics, you are the one person who is most likely to understand the issues that are responsible for the conflict. If it is happening with one person, then if may be an issue that is affecting the entire program. In that case, you really are the one person who should be involved.

Community

Someone will need to be the face of the Robotics Program to the Community. This should be either the Robotics leadership or a physician who is considered your top surgeon performing robotic assisted surgery. Work closely with your hospital's Marketing and Public Relations (PR) Departments. It is important that you have an "elevator speech" ready at a moments notice. It is not uncommon for the media to contact PR with a pressing story and need to talk to a physician. You or your designee want to always be available for that opportunity. No one is asking you to make-up stuff. Be authentic.

Speak positively about the strengths of your program. If there is a controversial point, then address it. New studies come out all the time in healthcare innovation. Remember to keep up on the literature, and be able to speak to it when asked. This can also be a great topic of conversation at your committee meetings. It will keep your team engaged and informed.

Robotics Committee

This is the group that will be your biggest challenge. The reason is because it is usually composed of members of all the other stakeholder groups. There will always be competing priorities and agendas in the room even in the most organized programs. Identify a common agenda for the team during your first meetings. Data is usually an item of interest for everyone. You will want to know what data you can track, what data you need to track and how your team will obtain access to a reliable data source. Risk adjusted and bench-marked data is always nice to have but not essential in the early days of a program. As your program grows, access to the robot will become an issue so utilization metrics will become important. With growth will come issues of quality and cost, so make sure your program plans on having access to this type of data when the time comes. As you can imagine, if you are going to be having discussion around quality, then risk adjusted data will become very important to your physicians. Be their advocate. They will not let this issue die, and neither should you.

One last bit of advice I've adapted from a column titled "Leader Time: 8 things every CEO must know (from memory)" written by Peter DeMarco.[8]

1) **People:** Know everyone by name on your Robotics Committee and know or be familiar with every surgeon and nurse working in the robotics room.

---

[8] DeMarco, Peter. Leader Time: 8 things every CEO must know (from memory). The Business Journals. March 1, 2016.

2) **Prices:** What are you charging for your services? Is it competitive with the system down the road? You may not think this is important, but once you become a leader people will expect you to, at least, have some idea of this.
3) **Pay:** It is probably not important to know the exact compensation package everyone on your team has, but it is important for you to have some idea if your rates are competitive in the market. The last thing you want to do is to train a robotics team and have them jump to the competition because the hospital did not have a competitive compensation and benefits package.
4) **Purchasing:** This includes the equipment, instruments, accessories and supplies. You don't need know how much each Band-Aid costs but you should be familiar with all of the higher priced items. Also, have some idea of what alternative supplies cost such as reprocessed equipment when applicable.
5) **Presentation:** Be familiar with how your program is being presented to the community through your website or social media. Check out your website yourself, and visit your program's Facebook or LinkedIn pages, assuming your program has either of these. If so, then use them to your advantage. If not, then consider coordinating with marketing on setting up these web pages.
6) **Perception:** How is your program being perceived by your stakeholders? A great way to do this is to ask friends, family and people in your community. Talk to patients and find out why they choose your program.
7) **Performance:** You must be familiar with your metrics and should know what your current dashboard shows. You will also want to know about significant changes in your metrics. When you see improvement, then feel confident and brag about it. Your team expects you to do this. Trumpet the team's success. It is a big booster for morale.

8) **Priorities:** The program's priorities will change as the program matures. This is normal. You will have to constantly be reassessing your priorities with guidance from your stakeholder groups.

CHAPTER 8
PERFORMANCE IMPROVEMENT

Performance Improvement (PI) is the primary goal of all programs. If it were easy, then everyone would be doing it effectively. There are many different ways to approach PI. For the purposes of this book, it is up to the individual program to decide which specific approach to use. Many, if not all, healthcare systems already have some type of approach established in their organization. What we intend to accomplish in this chapter is to focus on those processes and metrics that appear to drive exceptional performance in hospitals that utilize robotic assisted surgery. In doing so, we will organize the processes based on the unit in which they will occur and address metrics separately.

SYSTEM LEVEL

The most effective way to approach a problem is to address it at the level where the source of the problem exists. If we observe a problem at a facility and realize the problem is in a software product that is at the enterprise level, then trying to fix it at the facility is probably not going to work, and if it does, then it will probably still be a problem at other facilities in the system. When we work at a facility, it is not always easy to see how we are really part of a larger system. This is true even for a hospital that is not part of a larger company. In this case they can be part of a city, county, state and national healthcare system. For more practical purposes, when we think of a System Robotics Program, we are usually referring to one that has two or more facilities, with two or more robots.

The most common question from facilities that are part of larger systems is, "Why do we need a System Robotics Program?" The reason our system did it is because the data told us we needed to do it. The data demonstrated significant variation across our system. It was how we discovered that facilities that had highly organized Robotics Committees were performing much better than facilities that lacked an organized approach to their program. While the infrastructure needs are basically the same for a systems team, there are some subtle differences between a facility based Robotics Committee and that of the system.

1) The most obvious first question to ask is "Do you have a System Robotics Committee?" Having one which closely follows the structure of your facilities is the first step in being able to address problems at the system.
2) Are you tracking performance at the system level and breaking is down by facility? This allows you to see the range of variation across your system. It also creates the ability to identify areas of outstanding performance, or positive outliers. Understanding

what this group of robotics teams are doing will enable the system to distribute this knowledge across the system.

3) Are you able to track the performance of your busiest surgeons as well as identify which surgeons are obtaining exceptional outcomes? What these groups are doing as part of their practice will be an incredible source of information. Remember, there are many ways to do something that leads to poor performance, but only a few ways that consistently leads to outstanding performance. You want to understand what these individuals are doing as well as the teams that support them.

4) Do you have an organized approach to providing feedback to your teams and surgeons? You cannot expect anything or anyone to improve without feedback on performance?

5) Does your system utilize standardized criteria for credentialing, privileges and reappointment? This tends to be one of the more difficult elements for systems to achieve. More often than not, this is determined by the individual bylaws of the hospital's Medical Staff and there tends to be quite a difference of opinion on how this should be set.

6) Does your program work with your systems Value Analysis Program? A System Robotics Committee should be considered the subject matter experts on robotics for your system. Any discussions that are occurring with regards to value analysis should be including this group.

OPERATING ROOM

One of the biggest mistakes a hospital can make is to buy a robot, park it in the operating room, sit back passively and hope it performs miracles. As they say, hope is not a strategy. A robot is not only a tool to assist in surgery, it is a fixed asset. Investopedia defines an asset as a "resource with economic value that a company owns or controls with the expectation that it will provide future benefit.[9]"

---

[9] http://www.investopedia.com/terms/a/asset.asp

Assets need to be used, serviced and most importantly managed in order to be optimally productive. If you think robots are expensive, then just try to own one that is not being used, or not being managed for efficiency and effectiveness. At the facility level, managing robots begins in the operating room, which is where you should also focus your performance improvement at the start of the program.

1) Do you have a dedicated robotics team for the majority of the cases being performed in your OR? Just like volume matters for a surgeon, it also matters for an OR team. Inexperienced teams are costlier in the long run than the cost it requires to train and maintain a regular team for your robotic cases. Not only does it pay off in terms of utilization, the effect it will have on your team's satisfaction alone is worth it. Happy teams are productive teams and productive teams are good for hospitals and the patients they serve.

2) Do your teams regularly review the preference cards with the surgeons? There is a phenomenon sometimes referred to as "preference card creep". It is not uncommon over the course of time for a surgeon who is trying out a new device to say, "Add this to my preference card." The problem with this is they may never or rarely use that device again, and yet it will be opened for each case as long as it remains on the preference card. This then falls into the category of open and unused, which is waste. Surgeons need to review their cards at least annually to make sure this is not happening. The "always, sometimes, never" approach is one way to review preference cards. If a surgeon always uses a supply item, then leave it on the card. If they say sometimes, then assure them it will be in the room, but not opened for the case unless requested. If they say never, then remove it.

3) Are you tracking your robotic utilization? This refers to the number of cases per year per robot. Excluding weekends and holidays, there are about 250 OR days in a year. If you staff the

robot for 8 hours a day, then calculating utilization is based on 2000 hours of capacity. If you average 2-three hour cases per day with up to an hour after each case for turnover, then the maximum number of annual cases is 500 per year. Targeting 80% of maximum is considered a safe number to use for optimal utilization. This works out to be 400 cases a year per robot. This will vary depending on the types of cases you are doing. Programs with a high volume of shorter cases can expect to do more and programs with a more complex case mix will do less.

4) Are you tracking your robotic room utilization? Most programs will designate a specific room to park their robot. Moving robots around impacts turnover and increases the risk of damage to the robot. It is important to know how often cases performed in the designated robot room are really robotic cases. One of the most common complaints among surgeons is access to the robot. It is not uncommon to find that the primary reason surgeons do not have access to the robot is because non-robotic cases are being scheduled in the room with the robot. OR managers are usually incentivized to maximize total OR capacity. In order to do this, they may find it necessary to book non-robotic cases in the room with the robot. Doing so makes it difficult for surgeons to get access to the robot. This is another good reason to have a Robotics Committee. This group can monitor this, and help develop rules around scheduling in the room with a robot. One solution to this is to reserve the room with the robot for robotic cases only. This Robotic block time can be lifted at an agreed upon time prior to the day of surgery. For example, if there are no cases scheduled or there is time still available in the room with the robot 5 business days or less prior to scheduling, then a non-robotic case can be scheduled. The goal here is to allow access to the robot and optimize total OR capacity.

5) Does each case type (hysterectomy, prostatectomy, low anterior resection, etc.) have a standardized in-room configuration? This creates efficiency for the team. There is no second guessing

based on which surgeon or anesthesiologist is doing the case. This takes some consensus agreement among the physicians, and the best place for these types of discussions are in your Robotics Committee, not in the pre-op holding area.

6) Does each case type have a standardized case cart? If you can get agreement on the in-room configuration, then the next step is to develop agreement on standardizing the case cart for each type of procedure. This is going to be much more difficult to do as surgeons tend to be very selective about the tools they use. Do not try to get everyone to standardize in one day. This will be more disruptive for you surgeons and your OR. Start with the "always, sometimes and never" approach previously described. This will eliminate much of the waste. Now that you can see what everyone must "always" use, see if they are willing to try or eliminate different instruments. The goal now becomes one of gradually moving toward a standardized case cart which is the most cost-effective supply solution for that case. This may take some time to do. There is a very important reason for this. Surgeons learn to use different instruments to perform the same operation. In that learning process, they become very comfortable and attached to those instruments. It is more than just attached, they feel a sense of mastery with that instrument, and that feeling affects their confidence in performing an operation. If the instruments are abruptly changed, for no apparent reason, then you will have a very unhappy surgeon. It takes time for a surgeon to develop a sense of comfort with an instrument. Give them the time, and develop a process for transitioning through discussion at your Robotics Committee.

7) Do you review charge levels for your cases, and are they appropriate for the case type? At one system it was "common knowledge" that the robot costs twice as much per minute than the average case. On a deeper dive into the true cost it was discovered that all robotic cases were routinely charged at the highest level. This was done irrespective of the complexity or

actual resources used during the case. This is not an accurate reflection of the true cost. Since the highest charge level is twice the cost of the average charge level, it should come as no surprise that it "cost more". It didn't. The charges were arbitrarily set there almost a decade ago, and no one bothered to change them or even check them.

8) Do you monitor and have rules about which types of cases can be performed on the robot? While it is understood that surgeons need time performing easier cases on the robot before progressing to more complicated cases, there is a limit. This is where your Robotics Committee needs to take action if this is an issue at your facility. A surgeon who routinely uses the robot to perform all diagnostic laparoscopies is adding cost to the case and negligible experience to their robotic skill set. For low volume surgeons who use simple cases to gain experience on the robot, a simulator is an excellent way to develop familiarity with the technology without putting a patient at risk or adding unnecessary cost.

9) Is the patient's final status determined in or prior to the recovery room most of the time? With the advent of minimally invasive surgery, cases have been transitioning from inpatient status to the outpatient or observation status. The difference in reimbursement is substantial. Unfortunately, cases that are preapproved for inpatient status and meet medical necessity can be placed in observation out of habit. It is important for your program to make sure patients are placed in the appropriate status, in order to insure it is billed correctly. It is also important for the surgeon's documentation for medical necessity, diagnostic and procedure coding and patient status are in alignment with the hospitals. This will help reduce any denials for both surgeon and hospital. The best place to get this determined is before the patient leaves the recovery room.

10) Do you track standard metrics for your cases and include them in your Robotics Committee dashboard? If so do, then do

you have annual goals for these metrics and action plans to achieve these goals? Later we will discuss the specific metrics. Keep in mind the old saying, "you manage what you measure". If you want to improve, then you will need to find a way to measure it and track it over time in order to sustain an effective performance improvement program.

MEDICAL/SURGICAL UNIT

1) Do you use clinical pathways for your more common cases? Clinical Pathways such as Enhanced Recovery after Surgery produce better outcomes. If you are not utilizing these types of pathways, then it is essential for your Robotics Committee or Department of Surgery to consider developing them. They are associated with lower complications, length of stay, readmissions and cost.
2) Do you have standardized criteria for discharging patients? Established criteria for discharging patients makes it a more predictable process for the nursing staff and patients. For many elective cases, especially when a clinical pathway is being used, standard discharge criteria become part of the clinical pathway.
3) Are patients provided either reading material or a video tutorial which educates them about their recovery and what to expect? Patients who are informed regarding their care and what to expect tend to have better overall satisfaction with their care.

SURGEON/TEAM

1) Do you have standardized credentialing, privileges and reappointment criteria for surgeons? Most hospitals will be able to respond yes to this. What this question is really getting at is, "do you know what this criterion is and what are the reasons for using it?" More mature programs tend to have a higher threshold for training requirements and higher number of minimal annual cases for initial appointment and reappointment.

2) Do your surgeons average more than 24 cases/year? There is nothing magical about this number except it was the least number of cases that separated the busiest surgeons from the least busy for the top 20th percentile. In that study, the busiest surgeons were averaging 80 case/year. The point is, high volume surgeons tend to have lower complication rates and lower costs.
3) Do your surgeons have access to a simulator and use it regularly? You can own a simulator, but if no one uses it, then it is just taking up space.
4) Does your robotics team use simulation for training? Teams that practice together perform better than if they don't.
5) Do new members of your robotics team go through a standardized training program prior to working independently for a robotic assisted case? Team members who have not been checked out on the technology will lead to poor utilization and higher cost.
6) Do you have established performance metrics for your surgeons and your team? In order for a team to improve, there must be a metric for them to assess performance.
7) Does your team get regular feedback on performance? Individuals and teams that get feedback on their performance do better than when they get no feedback. The more regular and immediate the feedback, the better.
8) Does your team have a say in what performance metrics they will use, and what will be the goals for the team? Allowing teams to have a say in how they will be evaluated promotes buy-in, and reinforces the message that they can control their environment. They are also more likely to be engaged in their work which is great for morale.

METRICS

If infrastructure is the foundation on which your Robotics Program will function, data is the foundation on which your Performance

Improvement Program for robotics will function. There are many different types of data and many different sources. George Box, the statistician, once quipped that "All models are wrong, but some are useful". The same could be said of data, in that all data is imperfect, but some data can be useful. When working with data there is a tendency to either completely ignore the data as irrelevant, or embrace it as the ultimate source of truth. All data, or at least all of the data I've encountered, is neither. There is one thing I've learned about data and that is this: How you use data is just as important as whether you use it.

The first thing to keep in mind is to accept your data for what it is. Don't sit around waiting for the perfect data source. This will only serve to paralyze your efforts at performance improvement. Instead of having massive amounts of data points, focus on a few that are consistent, reliable and worth the time and effort to improve. Data costs in terms of time, money and resources to collect, analyze and distribute. Make sure you are using only the most cost-effective metrics that will lead to real performance improvement. The following is a list of metrics to consider tracking for your program.

Utilization Metrics

1) **Volume**
    a. Cases/year: Facility and System (#)
    b. Cases/year: Specialty (# and %)
    c. Cases/year: Robot (# and % if more than one)
    d. Cases/year: Surgeon (# and %)
2) **Throughput**
    a. Accurate case duration estimate (%): When a surgeon schedules a case, how accurate is their case duration estimate +/- an accepted margin of error.
    b. First case on time/early start (%): OR's with a low percentage of starting on time for the first scheduled case

of the day may find it difficult to effectively manage an OR schedule.
c. Subsequent case on time/early (%): ditto to previous comment.
d. Patient in to incision time (minutes): How long does it take to provide anesthesia and prep/drape?
e. Patient close to out time (minutes): How long does it take to wake the patient up and move them out of the room?
f. Patient incision to patient close time (minutes): How long does it take to perform the operation?
g. Patient out to next patient in time (minutes): Also known as turnover time.
h. Preadmission screening (%): Patients with preadmission screening generally have a smoother flow through the OR.
i. Surgical checklist (%): Safe Surgery procedures are associated with a reduced likelihood of serious reportable events.
j. Robotic prime time utilization (%): Often compared to non-robotic utilization in the same room where the robot is placed.
k. Close to cut time (minutes): Often advocated by the robot manufacturer as this time is reportedly correlated with surgeon satisfaction. Surgeons love to operate, and longer down time between operating is considered dissatisfying to a surgeon.
l. Docking/undocking time (minutes): This should improve with practice. A good starting metric for new programs.
m. Console Time (minutes): Represents the time the surgeon is performing the robotically assisted portion of the case.

Cost Metrics

1) Average cost/case: All ($)
2) Average cost/case: Specialty ($)
3) Average cost/case: Case Type ($)

4) Average cost/case: Surgeon ($)
5) Average supply cost/case: All ($)
6) Average supply cost/case: Specialty ($)
7) Average supply cost/case: Case Type ($)
8) Average supply cost/case: Surgeon ($)
9) O/E cost/case: This is an observed to expected ratio for cost/case. A company (e.g. Premier, Advisory Board Crimson) that provides data analysis will typically report in this manner. These are risk-adjusted and bench-marked reports and are an excellent source of information regarding your program.
10) O/E HLOS/case: Same as above except for hospital length of stay. This can be used in conjunction with O/E cost/case. For example, if your cost/case is high and your HLOS/case is low or average, then look to supply cost as the driver of the total cost.

Quality Metrics

1) O/E complication ratio: This can be broken down by hospital, specialty, surgeon or case type.
2) O/E mortality ratio: The mortality rates for robotic assisted surgery tend to be very low, so unless you are a high volume facility this may not be a very helpful rate to follow. All mortalities should be followed up by your quality department and when applicable peer review.
3) O/E readmission ratio: This can be broken down by hospital, specialty, surgeon or case type.

It may take some time and some diligent data mining at first to know what metrics you want to track. Since hospitals tend to have different data sources, some data may be easier to obtain. Start with what you have and make the most of it. Don't let the lack of access to your organizations data source be an obstacle to your performance improvement efforts. If your case volume is relatively low, it may still be possible to track some of the data on a simple spreadsheet or even a dry erase board. As long as you are not tracking protected health information and are using the data for performance improvement you should be on safe ground. The best option for most hospitals is to include your quality department in this effort. They are always looking for PI projects, especially when they have an engaged team supporting them. Most of the quality departments will already have a model for performance improvement such as PDSA or DMAIC. If so, then piggyback on their resources to jump start your program. If not, then below is a suggested model to use for performance improvement.

Ten Steps to Best Practice

1) Establish a Baseline: Once you have determined the metrics you will follow, then establish a baseline period and know what your starting point is.
2) Benchmarking: If you have access to a program that will risk-adjust and bench-mark your data, then you are off to a great start. If you do not have access to such a resource, then don't despair. You can always contact resources such as AORN to find national averages for many of the previously mentioned metrics. In addition, with your baseline metrics you will now have your historical baseline as a comparison.
3) Identify Opportunity: Any gap between a benchmark best practice and current practice is a potential opportunity. Prioritize your opportunities, and select the top one to four for improvement projects.
4) Attribution Analysis: Determine the source of attribution for the opportunity. If all of the variation from best practice for a particular metric is from one source, then focus your efforts on that source. It's commonly thought that it is usually the surgeon who is the problem, when in fact, the data may tell you it is really a specialty or even system source.
5) EBM: Review the literature for evidence based medicine (EBM) of best practice or practice guidelines. If you notice that you have a higher than expected complication of ileus for your bowel surgery cases, then you will want to develop and implement an

enhanced recovery after surgery clinical pathway for these types of cases.

6) PBE: Perform data mining to identify practice based evidence (PBE) for best practice. There may be a physician or team in your system who is already achieving outstanding results. Find out what they are doing, and do it.

7) Plan & Do: Once you have identified, either with EBM or PBE, develop and implement a clinical or operational practice that can be tracked with specific process measures and correlated with the specific outcome of interest. If this is your first time doing this, then keep it simple. The more complicated the process and the larger the organization, the more difficult it will be to implement.

8) Process Measures: Measure before and after process measures to determine the effectiveness of the implementation. In rolling out an enhanced recovery program following bowel surgery across multiple hospitals in seven states, we decided to focus on just two process measures. These were early alimentation and early ambulation. By getting the system focused on these two measures we demonstrated a marked improvement in a very short period of time[10]. Please also note in the chart below, that we used process control charts. If you have a resource for creating these types of charts, then this is a strongly recommended approach. It allows you to determine if the process steps are really improving or not. It also allows you to track performance over time. As you can see in this example, this clinical practice was sustainable over many years and continues to this day.

---

1) [10] Loftus T, Stelton S, Efaw BW, Bloomstone J. A system wide care pathway for enhanced recovery after bowel surgery focusing on alimentation and ambulation reduces complications and readmissions. J. Healthcare Quality. Published online Feb 20 2014. doi: 10.1111/jhq.12068

9) Outcome Measures: Measure before and after outcomes to determine the effectiveness of the implementation on patient care. This is why we are, after all, doing this. For example, in our bowel surgery initiative the improvement in process measures (See Chart 1) was associated with a 28.8% reduction in complications and a 17.5% reduction in readmissions.
10) Standardize: Standardize to the new best practice and continue to support it until is is confirmed to be stabilized.

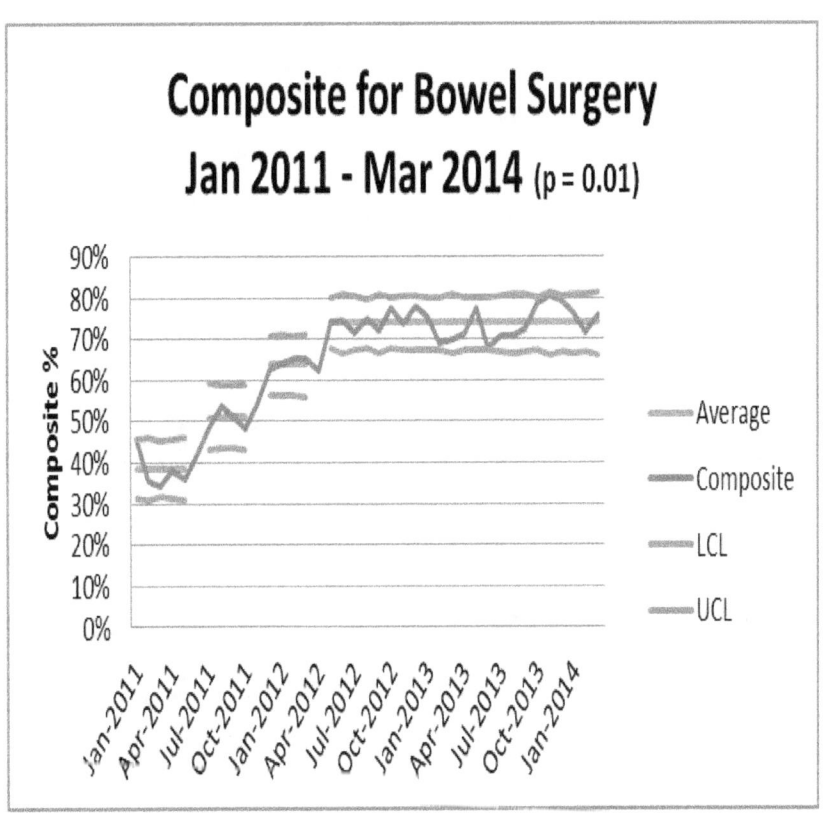

Chart 1: *Composite score for Enhanced Recovery after Bowel Surgery process steps (ambulation & alimentation) before and after system implementation (July 1, 2011) and through first quarter of 2014.*

APPENDIX
COMMITTEE CHECKLIST

By now you have probably figured out, this is not just a book on how to start-up a successful Robotics Program. It is also an introduction on how to start-up successful programs in general. Each program will have specific features which will distinguish it from other programs. The common features are core elements that every program will need to be successful. I've organized these core features into "Seven Pillars". After years of starting different types of programs, the presence of these seven pillars became an unmistakable pattern in the most successful programs. As you develop your Robotics Program look for their presence. If you are struggling, then re-examine the program, and see if you are adhering to the elements outlined in each chapter.

All successful outcomes are the result of a series of process steps. Some must occur before others, and there will be some that will not work in all environments. In this Appendix a recommended checklist will be provided. The goal is to take you through the planning, the launch of your first Robotics Committee meeting and establishing a performance improvement program. This whole process can be done in less than ninety days. A more aggressive timeline can be a done based on your system's priorities. This checklist will take you through the first three meetings. It should work if you plan on meeting every month, every other month or every quarter.

After establishing the Leads and Administrative Assistant, each step will be followed by a recommended responsible person. This will be in parenthesis and will be bolded. Attempt to achieve each item in the checklist, as best you can, in order to launch your program. During this process, keep track of what was essential and what was not. Check the box next to essential items and cross off the list those items found to be non-essential. For example, it may not be customary to provide refreshments at meetings in your organization, then cross this off the list. While the list may seem imposing at first, it is designed to help you develop good habits. It is these good habits that you will want to hardwire into your program. By doing so, you will hardwire barriers to disengaging behaviors. The more effectively you can do this, the more time you will have to focus on engaging your team and building a successful program.

The Robotics Committee will be the lead executive forum for robotics in your organization. Its role is essential for a high performing program. Setting it up correctly from the beginning will allow you reach a higher level of performance much earlier in the process. It requires a lot of "front end" work as they say. Regardless of whether your committee will be meeting every month, every other month or every quarter, plan on meeting at least once a month until you have developed a performance improvement project. The checklist is set up to get you there in three months. You may elect to do it quicker but really avoid taking any longer than three months. Create a sense of urgency and capitalize on the momentum to start-up your program. It will be worth it.

I wish you the best of luck in your program's success. If you have any comments or questions, then feel free to contact me at:

TerryLoftusMD@gmail.com or through

www.LoftusHealth.com

Planning for the Robotics Committee Meeting

- ☐ Identify a Physician Lead.
- ☐ Identify a Nursing or Administrative Lead.
- ☐ Identify an Administrative Assistant. **(AA)**
- ☐ Set up first planning meeting. **(AA)**
- ☐ Physician & Nursing Leads to meet with stakeholder leadership to determine acceptance of Robotics Program and which model (Medical Staff vs. Administrative) to use. **(Leads)**
- ☐ Stakeholder leadership list.
  - Chief of Staff
  - Department Chairs (Surgery, OB/Gyn, Urology, Cardiovascular, ENT, Anesthesia)
  - CEO of hospital
  - CMO of hospital
  - CNO of hospital
  - Perioperative Services Director
  - Supply Chain Services Director
- ☐ Confirm type of model to use (Medical Staff, Administrative vs. Hybrid). **(Leads)**
- ☐ Identify membership for Robotics Committee. **(Leads)**
- ☐ Robotics Committee membership list and contact information.
  - Physician Lead
  - Nursing Lead
  - Administrative Representative
  - Administrative Assistant
  - Robotics Coordinator
  - Quality Representative
  - Data Analyst
  - Supply Chain Representative
  - Specialty Representatives

- Cardiovascular Representative
- ENT Representative
- General Surgery Representative
- OB/Gyn Representative
- Orthopedics Representative
- Urology Representative

☐ Set up Prep meeting for first committee meeting. **(AA)**

Prep Meeting

☐ Establish calendar of meeting dates. **(Leads & AA)**
☐ Confirm meetings are consistent (e.g. first Thursday of the Month at 5:00 PM in Boardroom. **(Leads & AA)**
☐ Determine if teleconference capability needed. **(Leads & AA)**
☐ Develop list of membership with contact information. **(Leads & AA)**
☐ Develop meeting agenda template. **(Leads & AA)**
☐ Determine if any documents will be needed for first meeting. **(Leads & AA)**
☐ Determine purpose of committee. **(Leads)**
☐ Develop first meeting's agenda to include: **(Leads & AA)**
  - Name of committee
  - Purpose of meeting
  - Type of meeting
  - Date of meeting
  - Time of meeting
  - Location of meeting
  - Call in number with any code (if any)
  - Web link for remote conferencing (if any)
  - Members names and role for attendance
  - Note taker for meeting minutes
  - Time keeper
  - Note for any member preparation for meeting
  - Call meeting to order

- o Call to review previous meeting minutes
- o Action items
- o Responsible party for action items
- o Due date for action items
- o Agenda items
- o Presenter for agenda items
- o Time allotted for each item
- o New action items
- o Responsible party for new action items
- o Due date for new action items
- o Call for new agenda items for next meeting
- o Reminder of next meeting date, time location
- ☐ Confirm centralized repository for program documents.
- ☐ Confirm membership has access to centralized repository.

Prior to First Robotics Committee Meeting

- ☐ Confirm next meeting and teleconference capability. **(AA)**
- ☐ Confirm availability of audio-visual equipment and support contact information in case of problems. **(AA)**
- ☐ Send out next meeting notice to membership along with any specific instructions on how to access the meeting via phone or internet. **(AA)**
- ☐ Leadership to confirm final agenda. **(Leads)**
- ☐ Send out follow-up meeting notice along with final agenda and any documents to membership. **(AA)**
- ☐ Confirm RSVPs for meeting. **(AA)**
- ☐ Leadership to contact any membership non-responders to meeting notice to confirm RSVP. **(Leads)**
- ☐ Leadership to contact AA with any confirmed RSVP from non-responders. **(Leads)**
- ☐ Arrange refreshments (if any) for next meeting based on RSVP response. **(AA)**
- ☐ Prepare any paper documents needed for membership at the next meeting (agenda and documents). **(AA)**

First Robotics Committee Meeting

- ☐ Arrive 10 – 15 minutes before meeting begins. **(Leads & AA)**
- ☐ Set-up any audio-visual equipment and test. **(Leads or AA)**
- ☐ Set-up any teleconferencing equipment and test. **(Leads or AA)**
- ☐ Call in to any bridge line or web enabled teleconferencing 3 – 5 minutes before meeting begins. **(Leads or AA)**
- ☐ Remind call in members that meeting will begin promptly at designated time after making connection. **(Leads or AA)**
- ☐ Call meeting to order at designated time. **(Leads)**
- ☐ If meeting needs to begin later, notify members why meeting is delayed and expected start time. Repeat this every 2-3 minutes until meeting begins. **(Leads)**
- ☐ Take attendance. **(Leads or AA)**
- ☐ Review purpose of the meeting with membership. **(Leads)**
- ☐ Introduce new members and guests. **(Leads)**
- ☐ Remind members to disclose any pertinent potential conflict of interest during any meeting deliberations. **(Leads)**
- ☐ Work through agenda with intent to stick to allotted time for each agenda item. **(Leads)**
- ☐ If extra time needed for additional discussion of an agenda item, then ask members if it is okay to go past the allotted time. If so, then notify members of what the additional time limit will be. **(Leads)**
- ☐ Consider creating a Metrics Team to develop a list of potential metrics, and a mock-up of a dashboard for the next meeting. **(Leads)**
- ☐ Request any future agenda items to be forwarded to the Leads and the AA. **(Leads)**
- ☐ Announce next meeting date, time and location. **(Leads or AA)**
- ☐ Adjourn meeting on time. **(Leads)**

Prior to Second Robotics Committee Meeting

- ☐ Confirm next meeting and teleconference capability. **(AA)**
- ☐ Confirm availability of audio-visual equipment and support contact information in case of problems. **(AA)**
- ☐ Send out next meeting notice to membership along with any specific instructions on how to access the meeting via phone or internet. **(AA)**
- ☐ Schedule meeting with Metrics Team to determine potential metrics and mock-up of dashboard. **(AA)**
- ☐ Meet with Metrics team to finalize list of potential metrics. **(Leads)**
- ☐ Pull any data, graphs or charts needed for meeting and forward to Leads. **(Data Analyst)**
- ☐ Leadership to review any data, graphs, charts or documents for next meeting and forward to AA. **(Leads)**
- ☐ Leadership to develop, confirm and forward final agenda to AA. **(Leads)**
- ☐ Send out follow-up meeting notice along with final agenda and any documents to membership. **(AA)**
- ☐ Confirm RSVPs for meeting. **(AA)**
- ☐ AA to notify Leads of any non-responders to RSVP. **(AA)**
- ☐ Leadership to contact any membership non-responders to meeting notice to confirm RSVP. **(Leads)**
- ☐ Leadership to contact AA with any confirmed RSVP from non-responders. **(Leads)**
- ☐ Arrange refreshments (if any) for next meeting based on RSVP response. **(AA)**
- ☐ Prepare any paper documents needed for membership at the next meeting (agenda and documents). **(AA)**

Second Robotics Committee Meeting

- ☐ Arrive 10 – 15 minutes before meeting begins. **(Leads & AA)**
- ☐ Set-up any audio-visual equipment and test. **(Leads or AA)**
- ☐ Set-up any teleconferencing equipment and test. **(Leads or AA)**
- ☐ Call in to any bridge line or web enabled teleconferencing 3 – 5 minutes before meeting begins. **(Leads or AA)**
- ☐ Remind call in members that meeting will begin promptly at designated time after making connection. **(Leads or AA)**
- ☐ Call meeting to order at designated time. **(Leads)**
- ☐ If meeting needs to begin later, notify members why meeting is delayed and expected start time. Repeat this every 2-3 minutes until meeting begins. **(Leads)**
- ☐ Take attendance. **(Leads or AA)**
- ☐ Review purpose of the meeting with membership. **(Leads)**
- ☐ Introduce new members and guests. **(Leads)**
- ☐ Remind members to disclose any pertinent potential conflict of interest during any meeting deliberations. **(Leads)**
- ☐ Review previous meeting minutes. **(Membership)**
- ☐ Approve previous meeting minutes. **(Membership)**
- ☐ Work through agenda with intent to stick to allotted time for each agenda item. **(Leads)**
- ☐ If extra time needed for additional discussion of an agenda item, then ask members if it is okay to go past the allotted time. If so, then notify members of what the additional time limit will be. **(Leads)**
- ☐ Review potential metrics and dashboard with membership. **(Leads)**
- ☐ Decide on which metrics to track. **(Membership)**
- ☐ Request any future agenda items to be forwarded to the Leads and the AA. **(Leads)**
- ☐ Announce next meeting date, time and location. **(Leads or AA)**

- ☐ Adjourn meeting on time. **(Leads)**

Prior to Third Robotics Committee Meeting

- ☐ Confirm next meeting and teleconference capability. **(AA)**
- ☐ Confirm availability of audio-visual equipment and support contact information in case of problems. **(AA)**
- ☐ Send out next meeting notice to membership along with any specific instructions on how to access the meeting via phone or internet. **(AA)**
- ☐ Schedule meeting for Leads with Data Analyst to review data. **(AA)**
- ☐ Meet with Data Analyst to identify opportunities for improvement. **(Leads)**
- ☐ Pull any data, graphs or charts needed for meeting and forward to Leads. **(Data Analyst)**
- ☐ Leadership to review any data, graphs, charts or documents for next meeting and forward to AA. **(Leads)**
- ☐ Leadership to develop a list of 3 – 5 goals for the Robotics Committee based on opportunities identified in data. **(Leads)**
- ☐ Leadership to develop a draft of how the committee will communicate its progress to its stakeholders. This can be in the form of a newsletter, memo or presentation. **(Leads)**
- ☐ Leadership to develop a communication plan with timeline and list of stakeholders for communication. **(Leads)**
- ☐ Leadership to develop, confirm and forward final agenda to AA. **(Leads)**
- ☐ Send out follow-up meeting notice along with final agenda and any documents to membership. **(AA)**
- ☐ Confirm RSVPs for meeting. **(AA)**
- ☐ AA to notify Leads of any non-responders to RSVP. **(AA)**
- ☐ Leadership to contact any membership non-responders to meeting notice to confirm RSVP. **(Leads)**
- ☐ Leadership to contact AA with any confirmed RSVP from non-responders. **(Leads)**

- ☐ Arrange refreshments (if any) for next meeting based on RSVP response. **(AA)**
- ☐ Prepare any paper documents needed for membership at the next meeting (agenda and documents). **(AA)**

Third Robotics Committee Meeting

- ☐ Arrive 10 – 15 minutes before meeting begins. **(Leads & AA)**
- ☐ Set-up any audio-visual equipment and test. **(Leads or AA)**
- ☐ Set-up any teleconferencing equipment and test. **(Leads or AA)**
- ☐ Call in to any bridge line or web enabled teleconferencing 3 – 5 minutes before meeting begins. **(Leads or AA)**
- ☐ Remind call in members that meeting will begin promptly at designated time after making connection. **(Leads or AA)**
- ☐ Call meeting to order at designated time. **(Leads)**
- ☐ If meeting needs to begin later, notify members why meeting is delayed and expected start time. Repeat this every 2-3 minutes until meeting begins. **(Leads)**
- ☐ Take attendance. **(Leads or AA)**
- ☐ Review purpose of the meeting with membership. **(Leads)**
- ☐ Introduce new members and guests. **(Leads)**
- ☐ Remind members to disclose any pertinent potential conflict of interest during any meeting deliberations. **(Leads)**
- ☐ Review previous meeting minutes. **(Membership)**
- ☐ Approve previous meeting minutes. **(Membership)**
- ☐ Work through agenda with intent to stick to allotted time for each agenda item. **(Leads)**
- ☐ If extra time needed for additional discussion of an agenda item, then ask members if it is okay to go past the allotted time. If so, then notify members of what the additional time limit will be. **(Leads)**
- ☐ Review metrics and dashboard with membership. **(Data Analyst)**

- ☐ Discuss opportunities for performance improvement with membership. **(Leads)**
- ☐ Propose and choose 1 – 3 opportunities for the committee to use for their first performance improvement project. **(Membership)**
- ☐ Decide on action plans or create teams to develop action plans for performance improvement projects. **(Membership)**
- ☐ Assign Leads and teams to each performance improvement project. **(Leads)**
- ☐ Decide on which metrics to track for performance improvement project(s) and SMART goals. **(Membership)**
- ☐ If needed, then review what "SMART" goals are with membership. **(Leads)**
- ☐ Present communication plan to membership for comments and approval. **(Leads)**
- ☐ Request any future agenda items to be forwarded to the Leads and the AA. **(Leads)**
- ☐ Announce next meeting date, time and location. **(Leads or AA)**
- ☐ Adjourn meeting on time. **(Leads)**

ABOUT THE AUTHOR

Dr. Terrence Loftus is the President of Loftus Health, a healthcare consulting company committed to educating and coaching the next generation of healthcare leadership on how to improve the delivery of healthcare. Prior to this, Dr. Loftus was the Medical Director of Surgical Services & Clinical Resources for Banner Health in Phoenix, Arizona. He obtained a BS in Psychology and an MBA from Arizona State University, and his Medical Degree is from the University of Arizona. He completed a residency in General Surgery at the University of Utah and a Trauma Surgery and Surgical Critical Care Fellowship at the University of Maryland's R Adams Cowley Shock Trauma Center in Baltimore, Maryland. Dr. Loftus is also a graduate of the Advanced Training Program for Executives and Quality Improvement Leaders sponsored by Intermountain Healthcare's Institute for Healthcare Delivery Research. Dr. Loftus has served in various leadership roles including Chief Medical Officer, as well as a Medical Director of a Surgical Intensive Care and a Level 1 Trauma Center. He is board certified in General Surgery and Surgical Critical Care, and is a Fellow in the American College of Surgery. He can be contacted by visiting his website at www.LoftusHealth.com.

www.ingramcontent.com/pod-product-compliance
Lightning Source LLC
Chambersburg PA
CBHW080935170526
45158CB00008B/2301